D1130432

The Brain Behind Pain

The Brain Behind Pain

Exploring the Mind-Body Connection

Akhtar Purvez, MD

ROWMAN & LITTLEFIELD
Lanham • Boulder • New York • London

Published by Rowman & Littlefield
An imprint of The Rowman & Littlefield Publishing Group, Inc.
4501 Forbes Boulevard, Suite 200, Lanham, Maryland 20706
www.rowman.com

86-90 Paul Street, London EC2A 4NE

Copyright © 2022 by Akhtar Purvez

All rights reserved. No part of this book may be reproduced in any form or by any electronic or mechanical means, including information storage and retrieval systems, without written permission from the publisher, except by a reviewer who may quote passages in a review.

British Library Cataloguing in Publication Information Available

Library of Congress Cataloging-in-Publication Data

Names: Purvez, Akhtar, author.
Title: The brain behind pain : exploring the mind-body connection / Akhtar Purvez, MD.
Description: Lanham : Rowman & Littlefield Publishers, [2022] | Includes bibliographical references and index.
Identifiers: LCCN 2022023334 (print) | LCCN 2022023335 (ebook) | ISBN 9781538172803 (cloth) | ISBN 9781538172810 (ebook)
Subjects: LCSH: Pain—Physiological aspects. | Pain—Psychological aspects. | Chronic pain. | Affective neuroscience.
Classification: LCC QP401 .P87 2022 (print) | LCC QP401 (ebook) | DDC 612.8—dc23/eng/20220720
LC record available at https://lccn.loc.gov/2022023334
LC ebook record available at https://lccn.loc.gov/2022023335

∞™ The paper used in this publication meets the minimum requirements of American National Standard for Information Sciences—Permanence of Paper for Printed Library Materials, ANSI/NISO Z39.48-1992.

To my mother and to the memory of my father.

By living an exemplary life, they inculcated me with the virtues of kindness, selfless service, and love of the fellow human being.

Contents

Preface

Yeli asi daag tacchun hyot maagan,
sontus di yut asi naad.
Saam dilan heth, wash narian kod,
gulshan kor aabad

When winter tended to gnash our wounds,
we evoked spring.
Assayed our hearts, flexed our arms
brought the garden to bloom.

—*Wize' Wize'* by Muzaffar Aazim, Kashmiri poet and scholar

Welcome to the new frontier in humanity's battle against pain. We have met the enemy, as the saying goes, and he is us. In other words, we have learned that pain is not some malevolent outside force, but rather a product of our own inner circuitry. And we're discovering more about that every day.

As a physician specialized in treating pain, I find this revelation both inspiring and intimidating. It is inspiring because it holds out the hope of help for so many people; it's intimidating because every step forward brings with it a host of new questions and challenges.

Knowledge is power, we are told, and that statement is hard to refute. Yet the act of knowing can also bring a measure of comfort, for what we most fear is usually what we least understand.

That's why I'm writing this book, because pain remains one of life's great puzzles—not just for the average person, but for the scientific and medical community at large. At the same time, the pursuit of that mystery has pried open yet another door into the endless wonders of the human body—wonders that most of us never think about.

I happen to focus on pain in my practice, but it is an unwelcome gate-crasher in virtually every other medical specialty. Be it orthopedics, neurosurgery, oncology, neurology, internal medicine, obstetrics, gynecology, or any other healing discipline, the question of how much a condition or procedure is likely to hurt is often one of the first things that most patients want to know.

Which only makes sense. Part of that arises from our natural fear of pain; it is also one of the only indicators we have of the status of an injury or illness. Most of us don't ask for medical help unless something hurts us. Then, once we began taking medication or following some other advice from a doctor, we expect the pain to quickly and steadily recede.

That doesn't always happen with some of my patients, though, because chronic pain has a way of reaching a plateau and stubbornly defying all efforts to make it go away. For health-care providers who have come to rely heavily on their ever-evolving diagnostic tools, this can create a crisis of confidence.

What can we make of long-healed injuries that still hurt or of pain that seems to attack from out of nowhere? In some cases, we can stare at an MRI all day without getting answers.

Most doctors hate to say, "I don't know," and sometimes their reactions to these chronic pain anomalies have not reflected well on their profession. Rather than admitting that they didn't have all the answers, some responded by blaming their patients. "It's all in your head," they told them.

Eventually, we discovered that was actually true.

Many afflictions of the human body are relatively simple to explain and combat. With chronic pain—and, indeed, pain in general—conducting an investigation into its cause can be like wading into a swamp.

Back in the 1970s, medicine was partly represented in popular culture by a TV program called *Marcus Welby, M.D*. Dr. Welby, portrayed by Robert Young, seemed to deal with only one patient a week. He always spent considerable time consulting with that person, sometimes over coffee in the patient's home, listening intently to his or her symptoms and life issues. If he ever sent out a bill, it wasn't mentioned.

It's doubtful that such a benevolent and available healer ever existed in real life, and especially not now. Even before the COVID-19 pandemic swept in like a tidal wave, today's physicians were dealing with a severe time crunch.

"One day, everyone will be famous for fifteen minutes," the artist Andy Warhol once predicted. As it turns out, that is now the average time allotted for the physician to spend with each patient.

Forget leisurely coffee in a patient's kitchen—doctors today are lucky if they can grab a cup on the run from the nearest vending machine.

What has evolved is a no-win situation for any physician with an office practice. Those who are gently shooed out after the suggested fifteen minutes frequently feel shortchanged, but if an interaction takes longer, it throws a tightly scheduled assembly line off its track for the rest of the day. I can't even imagine how Dr. Welby would have dealt with chronic pain cases.

For those of us on the medical front lines, each pain-stricken visitor is merely a blip in a long day. We pull out an electronic device, transmit a prescription to their pharmacy, and perhaps never think about that person again until his or her next visit or anxious phone call. We are compartmentalized out of necessity, and each patient leaves that compartment as he or she walks out of the office door.

The problem is that pain generally follows that patient home, hitching a free ride on a compromised body. Then, once there, it extends its tentacles into every area of life.

Dr. Peter Abaci, author of *Take Charge of Your Chronic Pain*, was taught firsthand about the ubiquitous and all-consuming nature of pain when he tried to put a knee operation behind him.

"Getting from place to place was a struggle," he wrote. "I felt trapped in my body. Climbing up and down the stairs in my house exacerbated the pain. Even simply watching the nightly news, linked to the parade of disastrous events and dire predictions, seemed to make my knee hurt more. I felt like my entire life was unraveling."[1]

So debilitating can this be, both physically and mentally, that some chronic pain patients may choose suicide as the ultimate means of freeing themselves. Or they risk a fatal overdose by misusing medications that were prescribed to help them.

In reality, though, pain is not our enemy, but a necessary function of the very wise human body. Although that knowledge might be comforting in the abstract, it is small consolation to someone in the throes of a sleepless, pain-invaded night.

Humankind has tried to address the issue of pain for centuries, an effort that can perhaps be divided into a series of general periods.

First, there was the age of confusion. It began with the ancient Greek philosophers, whose theories as to the cause of pain were often contradictory, arrived at not by scientific research but by philosophical contemplation. A few who combined deep thoughts with actual medical experience, like Hippocrates and Galen, flirted with the truth, but their findings were arrived at in isolation and not generally disseminated at the time they were revealed.

After that came what might be called "the age of looking inward." Beginning with Rene Descartes in the 1660s and stretching until the work of Charles Scott Sherrington in the early twentieth century, this was a time when more and more scientists began searching the inscrutable depths of the human brain for clues to pain's origins. Sherrington's discovery of nociceptors was critical to this effort, and immediately preceded "the great counterattack," which focused more on relieving pain than delving into its origins.

This third age also brought a new player into the fight—multinational pharmaceutical companies, their appetite for wealth whetted by the German firm Bayer's discovery of aspirin in the 1890s. In quick succession, most of the analgesic pain relievers we know today made their appearance as other companies rushed to get in on the action. Moreover, ancient remedies such as opium derivatives were reinvented and redistributed.

Unfortunately, the profit motive soon overrode the simple quest for knowledge. Rather than pooling their well-financed resources, the drug companies quickly fell into fierce competition with each other. Even worse, drugs such as heroin and OxyContin eventually sparked the interest of another multinational force—the worldwide trade in illegal substances.

Thus the safety of desperate people fleeing from pain was compromised from above (the medical profession) and below (the purveyors of illicit narcotics). Since they were not only legal but prescribed by doctors and pushed by a host of TV ads, opioids were a dream come true for drug traffickers. Not that these drugs were necessarily dangerous in themselves if taken as prescribed and under strict medical supervision. Indeed, opioids may still be the recommended option for relieving intense short-term pain in some very specific cases, such as that following surgery.

When prescribed in an open-ended fashion for long-term chronic pain, however, opioids sometimes prove more of a threat to health than a blessing. Their beneficial effects tend to fray over time, prompting users to swallow more and more pills to maintain the same truce with their unwelcome "guest." Moreover, the fact that their doctors had essentially handed them the responsibility for their own medication programs—as well as having prescribed an oversupply in many cases—ratcheted up the temptation to increase their regular doses in an unsafe manner.

Complicating the problem was that many individuals who discovered the euphoric qualities of opioids began seeking them out solely for that purpose. The most intense of these was fentanyl, which was often mixed into opioid cocktails—especially those to be injected—as a way of adding to the "kick."

The perils of heroin were already well known, but many potential users of opioids had no idea that fentanyl is more than fifty times stronger. Partially as a result, a record more than one hundred thousand Americans died of drug overdoses from May 2020 to April 2021, a stark figure washed into the background by the arrival of COVID-19.

Some of these tragedies stemmed from careless recreational use, but many others were simply attempts at pain relief gone wrong. In the May 6, 2021, issue of *SF Weekly*, Benjamin Schneider addressed how the opioid plague has laid waste to one American city.

"In terms of loss of life, San Francisco's drug overdose epidemic is several times more severe than COVID-19. In 2020, the city recorded 258 COVID-19 deaths and 713 drug overdose deaths. About two-thirds of those overdose deaths involved fentanyl. This year is on track to be even more deadly. In the first three months of 2021, the city saw 203 overdose deaths. For context, during that same time period, nine homicides were reported in the city."[2]

We'll talk more about this in chapter 7, including how some pharmaceutical companies such as Inys, McKesson, and Purdue may have contributed to this national tragedy. The bright side, however, is that momentum is beginning to shift again in the field of pain treatment, from classic pharmaceutical means to safer alternatives based on new evidence-based medicine.

During the tragic rise in opioid overuse, many pain centers around the country closed and pain research in other places was cut back. Everything seemed to be under control, based on the American philosophy that "more is better." Since the problem could be taken care of so easily with medication, why did we need to find out where, in that indecipherable maze of nerves and neurons, pain was actually coming from?

People with drinking problems sometimes speak of the "bell curve effect." It goes something like this: "If I feel this good after four drinks, imagine how good I'll feel after eight."

With alcohol, passing over the top of that curve can lead to all sorts of unpleasant repercussions. With opioids, it can lead to assuming the main role in a funeral service.

I'm excited that we're back on track. I've never believed in a strictly pill-based practice, partly because I can't follow every patient home from my office to make sure they are taking their medications responsibly. Rather, I'm in favor of pain relief methods that make patients active participants in their treatment, not just passive pill consumers. Opioids remain one of the least employed weapons in my arsenal, used only in rare circumstances in combination with other approaches.

It's true that people experiencing chronic pain are not always easy to deal with. Many have become frustrated and angry, understandably wary of having to negotiate nearly constant discomfort. Because I'm a specialist, quite a few have been referred to my office by other physicians who were unable to "fix" them. Some have then fired me as their doctor after a treatment I prescribed failed to achieve miraculous results in a day or two.

Despite such unpleasant encounters, I've learned to be understanding. Enduring constant pain could push even a saint into a foul mood.

Because I maintain a busy practice, I can't be a dedicated, full-time researcher seeking new clues to how pain works and how we can manage it without risking lives. Still, I am involved in part-time research and make it a point to keep up with the latest information on the subject, from new interventional and medication studies including ongoing work on addiction-free medications.

If nothing else, we have learned that the vast majority of those who come to us with various types of chronic pain are not simply looking for sympathy (or OxyContin). They need—and deserve—our help.

Much of the research into chronic pain has occurred during the last two decades, reflecting the medical profession's late acceptance of conditions such as fibromyalgia. That, too, is an important part of the pain story.

According to Dr. James Clark, a renowned medical researcher in Charlottesville, Virginia, research in the area of fibromyalgia and other pain conditions is exploding. At his reputed research center, he has conducted as many as thirteen fibromyalgia trials. He tracks ongoing research and reports that, as per ClinicalTrials.gov, a government website monitoring these trials, there are more than two hundred research studies looking into fibromyalgia alone.

Seeking new information translates into hope for those given this cross to bear.

When viewed through the lens of our present knowledge, pain always will be part of human existence. Nevertheless, the same brain that is responsible for perceiving and modifying it can also help us find ways to come to terms with it.

It never hurts to learn more.

Chapter 1

The Mystery

Almost all living creatures experience pain to some degree, but only humans ask why. And they continue to ask, even after centuries of struggling to attach sense to something that once appeared unknowable. Pain is omnipresent yet simultaneously elusive. At times it seems to come from outside the body, at other times from within. Although the immediate cause is often obvious—a dog bite, a burn—it can also begin unexpectedly and remain for a lifetime. Even now, it remains one of nature's wild cards. No wonder the earliest attempts at explaining this phenomenon tended to credit (or blame) divine intervention. Pain, like God, felt like an outside force, so why wouldn't the two be connected?

Religious leaders in many cultures saw this belief as an open door to opportunity. "According to most religions in ancient times, pain . . . was sent by God to punish or test a person. It was, therefore, something to be borne to the best of one's ability. It was often purposely not treated for fear of interfering with God's will."[1] Ancient pain, then, required the attention of a high priest, not a physician. Since its presence was often seen as punishment for some misbehavior or corrupting attitude, the best thing someone in pain could do was beg for forgiveness. If the pain remained, it meant that their plea for mercy was insufficient. At least this offered those in pain an active means of confronting it. It also provided fertile ground for individual guilt, a staple of early religion, while bypassing any consideration of pain as an entity unto itself.

By the time the philosophers of ancient Greece took up the issue of pain, however, the religious leaders were already losing their grip on it. Why should a small child be tortured by a painful condition when the child had done nothing wrong? Why did the arrival of pain so often seem too much—or not enough—to be appropriate to the circumstances? How could these inconsistencies come from a just and all-knowing God?

Taking care not to unduly antagonize the spiritual establishment, learned men such as Aristotle, Plato, and Socrates tried to nudge the conversation in a more naturalistic direction. Perhaps, they theorized, pain was generated by some bodily organ such as the heart or the liver. This occurred when that organ became adversely affected by an imbalance among the four "humors"—blood, phlegm, yellow bile, and black bile. Even a cursory reading of the Greek philosophers makes it obvious that their writings were often a form of brainstorming, advancing multiple possibilities that in some cases even contradicted one another.

One intellectual path, for example, seemed to reach a dead end at a point of inverse logic. Since it was evident that pain and unhappiness were often simultaneous, one branch of thought held that pain must be caused by unhappiness, instead of the other way around. To Aristotle, meanwhile, pain could best be categorized as an emotion. Like Plato before him, he saw pain and pleasure not as sensations but as "passions of the soul."[2]

This spawned a prevailing view—stubbornly lodged in human consciousness—forever fusing pain with pleasure in an endless yin and yang relationship. "For moral excellence is concerned with pleasures and pain," Aristotle wrote. "It is on account of the pleasure that we do bad things, and on account of the pain that we abstain from noble ones."[3]

This insight, at least, has remained all too accurate.

By and large, the actual origins of pain did not enter into these discussions. Given the meager information available at the time, the great minds of ancient Greece simply lacked the intellectual tools to take on that mystery. Instead, they focused on the human reaction to pain. "Throughout his corpus, Plato treats pain and pleasure as opposites. The thesis is intuitive, but what exactly does opposition here amount to? Perhaps the most natural interpretation is in terms of qualities of experience: the painfulness of pain is the opposite of the pleasantness of pleasure. But now how is the opposition of phenomenal qualities to be understood?"[4] Plato and Aristotle argued that there was little or no difference between the pain of a broken leg and the emotional devastation of a broken love affair. If it wasn't pleasant, it must be pain.

Other early societies placed their focus not on religious or philosophical theories but on how to deal with pain on an everyday basis. For this, they turned to the natural world—anthropological studies have shown evidence of opium poppies being cultivated and harvested as early as 5000 BC. Meanwhile, Hippocrates (460–370 BC) kept a beacon glowing for later scientists. "In the Hippocratic texts, 'analgesia' is related to 'anesthesia' for the first time, when it is pointed out that an unconscious patient is insensitive to pain. Hippocrates and his followers rationalized pain as a clinical variable and as a valuable diagnostic and prognostic tool. They used expressive and precise adjectives and well-defined characteristics of pain, such as location, duration, or relation to other symptoms, to elucidate a disease process."[5] Though Hippocrates is perhaps best known today for unveiling the Hippocratic oath ("First, do no harm"), he also hammered together the underpinnings of modern pain theory.

"Hippocrates was the first physician to rationalize pain and use it as a valuable diagnostic and prognostic tool. He considered pain in relation to the overall clinical picture of the patient and not as an isolated symptom. He used the clinical features of pain as important manifestations in the disease process and as a major tool to outline the prognosis and severity of an illness. Numerous citations throughout the Hippocratic texts testify to this. For example, Hippocrates noted the time when the pain starts, to elucidate the course of the pathogenic process: When two different, distinct pains occurred simultaneously, he related the most intense one to the more serious illness."[6] Having been involved with anesthesiology early in my medical career, I can relate to and feel gratitude for what Hippocrates brought forth so many centuries ago. And for the purpose of this book, one quote among many perhaps best points the way to later science and scholarship: "Man ought to know," Hippocrates wrote, "that from the brain and from the brain only arise our pleasures, joys, laughter and jesting, as well as our sorrows, griefs, pain and tears."[7] True, he still gave credence to the influence of the four humors and other theories later disproven. Linking pain and the brain, however, was groundbreaking.

Another Greek, Herophilus (335–280 BC), known as the father of anatomy, was among the first to pick up the research thread that Hippocrates had spun. Much of what he discovered as a pioneer in human dissection was related to the brain and the nervous system.[8] "Herophilus is believed to be one of the first to differentiate nerves from blood vessels and tendons and to realize that nerves convey neural impulses. Although Charles Bell and François Magendie have both claimed the honor of discovering that dorsal spinal roots mediate sensation and ventral spinal roots carry motor fibers, it is believed that ancient anatomists such as Herophilus and Erasistratus had already appreciated the separate neural pathways during their time.

"It is suggested by Rufus in *De anatomia partium hominis* that both of them knew of the existence of 'two kinds of nerves,' that could either induce 'voluntary motion' or are 'capable of sensation.' It was Herophilus who made the point that damage to motor nerves induced paralysis. Herophilus named the meninges and ventricles in the brain, appreciated the division between cerebellum and cerebrum and recognized that the brain was the 'seat of intellect.'"[9]

Then came Galen of Pergamon.

The son of a wealthy Greek merchant, Galen was well educated and given opportunities beyond the reach of most of his contemporaries. His father originally steered him toward philosophy or mathematics until he had a dream in which the god Asclepius told him that his son should study medicine. Three years later, his father died, leaving Galen independently wealthy at age nineteen.[10] Many young men in that position might have settled in for a life of luxury at the helm of a lucrative family business, but Galen's fierce curiosity trumped his fortunate circumstances. Thus driven, he traveled and studied widely throughout Greece, Crete, and Cyprus, finally landing at the great medical school of Alexandria, where he was exposed to the various schools of thought in medicine. "In 157, aged 28, he returned to Pergamon as physician to the gladiators of the High Priest of Asia, one of the most influential and wealthy men in Asia Minor. Galen claimed that the High Priest chose him over other physicians after he eviscerated an ape and challenged other physicians to repair the damage. When they refused, Galen performed the surgery himself."[11]

Through his work with the gladiators, "he learned the importance of diet, fitness, hygiene and preventive measures, as well as living anatomy, and the treatment of fractures and severe trauma, referring to their wounds as 'windows into the body.' Only five deaths among the gladiators occurred while he held the post, compared to sixty in his predecessor's time, a result that is in general ascribed to the attention he paid to their wounds. At the same time, he pursued studies in theoretical medicine and philosophy."[12]

Later, Galen was drawn to the intellectual and cultural mecca of Rome, becoming court physician to the emperors Marcus Aurelius and Commodus. During that period, he continued his explorations into anatomy by dissecting animals, mostly monkeys and pigs. Human autopsies were forbidden in the Roman Empire, although there were rumors that Galen occasionally had his students secretly provide him with corpses of slain gladiators.[13]

In short, this was someone ideally suited to push the limits of human knowledge—well educated, well traveled, and with the time and resources to satisfy his thirst for learning. Galen also proved unafraid to confront what he considered to be the erroneous conclusions of his contemporaries.

"In the field of Neurosciences Galen was an eminent authority, who radiated through centuries. He traced new ways in brain anatomy and physiology, describing also numerous neurological disorders, with emphasis on the structural and functional background of the diseases and the etiology of the pathological processes. Galen characterized his scientific methods as proving . . . that they are based on objective, detailed observation and sufficient relevant experimental verification.

"Galen supported the encephalocentric theory of the human body, in contradiction to the cardio-centric aspect of Aristotle and the Stoics. According to Galen the brain is the 'hegemonicon,' the principal organ which rules and controls all the functions and the activities of the human body. The brain is the organ of cognition, volition, memory, fantasy, sensation, emotion, thinking, understanding and motor control. Brain is the seat of intelligence, the organ of the psychic expression."[14] Galen also followed Hippocrates on the necessity for professionalism and careful record keeping among physicians. Moreover, he had enough hands-on experience in patient care to understand an issue that still bedevils his successors today—the inherent subjectivity of pain as it is described by someone suffering from it. Long before the use of "scorecards" for pain, Galen attempted to find a common language for patients and physicians.

"His pain vocabulary, developed through a combination of consensus between patients and physicians' expert descriptions of their own pain, promises to link terminology univocally to sensation, turning patients' testimony about their subjective experience of pain into universally applicable diagnostic guidance."[15] Among his more questionable legacies, however, Galen also popularized the use of opium to fight pain, devising a method of dissolving it in wine for easier ingestion, and is blamed by some historians for instigating the addiction of Marcus Aurelius. Nevertheless, the influence of this man was enormous, in large part due to his prodigious productivity—twenty-one full-length works of more than one thousand pages. To use a twenty-first-century analogy, he essentially filled the search engines.

Perhaps the works of Galen and Herophilus might have been equally regarded, but most of those by the latter were lost in a fire at the library in Alexandria, leaving Galen the preeminent authority. Prior to the invention of the printing press in the mid-fifteenth century, extensive writings such as Galen's were copied by hand or with primitive stamps and woodblocks. Thus, they were not immediately available to every university or library and often did not make their appearance until long after their author had died. This meant that scholars could be dutifully exploring a topic without knowing that an earlier researcher had already plowed that ground.

The result was a decided lack of cooperative scholarship; new information that emerged was usually credited to an individual who was no longer around to defend it. That information, rather than being considered fact, was treated as belonging to Plato or Aristotle or Galen or whoever put his name on it. Often, efforts to refute those findings become more a matter of personal rivalry and jealousy than the desire to advance learning. Notwithstanding, much of Galen's writing eventually found its way around the then-known world. Some of it spread to the Moorish part of what is now Spain, prompting Islamic physicians such as Avicenna and al-Zahrawi to build on what they had read.

Just as Galen served under two Roman emperors, al-Zahrawi was court physician to the Andalusian caliph Al-Hakam II.[16] "He devoted his entire life and genius to the advancement of medicine as a whole and surgery in particular. He is known to have performed surgical treatments of head injuries, skull fractures, spinal injuries, hydrocephalus, subdural effusions and headache."[17] Avicenna fleshed out some of Galen's hypotheses and took issue with others. "Galen insisted that injuries (breach of continuity) were the only cause of pain. In contrast, Avicenna suggested that the true cause of pain was a change of the physical condition (temperament change) of the organ whether there was an injury present or not. Avicenna extended Galen's descriptions of 4 to 15 types of pain and used a terminology that is remarkably similar to that used in the McGill Pain Questionnaire."[18]

Although Galen, Herophilus, and al-Zahrawi all represented bright points of light during a dark period of intellectual dormancy, the true solution to the puzzle of pain dangled just out of reach. The idea that pain came from the brain was certainly a step in the right direction, but a limited one. There was still no explanation as to exactly how this occurred or how it could be addressed and perhaps influenced by the hand of a physician. It would be unfair, however, to think of members of the early medical community as intellectually inferior to those who came later. Indeed, the fact that they reached some of the conclusions they did is even more remarkable given the meager tools at their disposal. They were able to look at the human body only in its life-sized (and usually deceased) form, unaware of the microscopic elements that were actually driving the natural processes. Even the presence of germs was not totally accepted until centuries later.

Research during those intervening years was also hampered by the opposition of the powerful Catholic Church, not only to human dissections but to any medical thought that emphasized the physical rather than the spiritual aspects of humanity. This was exacerbated by the horrific reign of the Black Death.[19] Thus, after al-Zahrawi and some of his contemporaries briefly took the stage around the year 1000, more than four hundred years passed before the search for the origins of pain jump-started again, this time by Andreas Vesalius.

During the mid-1500s, Vesalius did much to break the stranglehold of Galen on contemporary medical theory by disproving some of his most cherished propositions. As might be expected, acceptance of this new direction was not universal. "Vesalius was aware of the value of his work, and of the academic jealousies that could work against him. He chose not to use a Venetian printer who was producing a rival, Galen-based anatomical tome, perhaps because he feared that the printer would deliberately delay the publication of his own study. Instead, he crossed the Alps to Basel, Switzerland, where, still fearing intellectual theft, he stayed to oversee the printing. He whiled away his time by boiling the body of an executed murderer to get at his bones. The reassembled skeleton is still displayed in the city's university."[20] Still, profound insights into the nature and cause of pain pop up only occasionally on the long timeline of brain research from 1500 to 1800. Prior to the nineteenth century, even the phenomenon of color blindness received more attention. One notable exception was the work of French scientist and philosopher Rene Descartes, who experienced a "light bulb moment" when exploring the phenomenon of phantom limb pain. Since there was no limb present to generate that pain, he realized that the impulse must be coming from the brain. Thus, he concluded, pain should be seen as "a perception that exists in the brain and makes the distinction between the neural phenomenon of sensory transduction . . . and the perceptual experience of pain." Blessed with diverse schooling in his youth—everything from music to mathematics to metaphysics—Descartes took a much more reality-based approach than most of his predecessors to the question of pain's origins. "Descartes is considered by many to be the father of modern philosophy, because his ideas departed widely from current understanding in the early 17th century, which was more feeling-based. While elements of his philosophy weren't completely new, his approach to them was. Descartes believed in basically clearing everything off the table, all preconceived and inherited notions, and starting fresh, putting back one by one the things that were certain, which for him began with the statement 'I exist.' From this sprang his most famous quote: 'I think; therefore I am.'"[21]

At the same time, Descartes knew that pain was not a product of his imagination, but an unmistakably natural process. Despite inserting vague references to the role of the soul in the origins of pain (largely to placate the Catholic Church, according to at least one biographer), the Frenchman was, at heart, a bedrock realist. "Descartes' manuscript, *Treatise of Man* . . . describes pain as a perception that exists in the brain and makes the distinction between the neural phenomenon of sensory transduction (today, known as nociception) and the perceptual experience of pain. What is essential to the development of Descartes' theory is his description of nerves, which he perceived as hollow tubules that convey both sensory and motor information."[22] Now that the writings of free thinkers like Descartes were able to be disseminated more extensively, it became obvious that the road to total understanding of pain was not a paved highway, but a series of widely separated stepping-stones across a lake of ignorance. "Descartes' work marked a major milestone in pain research and application. He created controversy in the world of pain research, which contributed to more debate and, ultimately, progress. His research at that time was revolutionary, especially when you consider the level of technology available."[23] The lack of serious research during the early years of medicine was hardly surprising. Certainly, it was no accident that both Galen and al-Zahrawi held high-profile (and, no doubt, well-paying) positions ministering to those at the top of the power structure. These fortunate circumstances undoubtedly provided the resources and the opportunities necessary for them to pry into the mysteries that attracted them. Just as rulers often became patrons of the arts, so some were patrons of science. Otherwise, there were few tangible rewards for making scientific discoveries beyond ego gratification. No pharmaceutical firms with deep pockets were standing by, no government grants awaited, no Nobel Prize. The average physician would not have had the time for extracurricular investigations.

Beyond that, only a small percentage of the population had more than a rudimentary education. Thus, whatever books may have been written about theories and discoveries were read by only a handful of other academicians, not an eager public audience seeking medical miracles. That left the universities, which is where many early scientists found a home. If nothing else, their students often provided a willing (and free) source of apprentices and research assistants. When you look at a published scientific paper today, you'll usually find multiple authors and several universities, often crossing international borders. At the time Descartes began making his mark, however, there was much more competition within the scientific community than collaboration. Often, instead of building on what had already been done, new researchers tried to tear it down.

After declaring a desire to place humans on the moon, twentieth-century man accomplished that goal in less than a decade. Finding the true cause of human pain—of far more interest for the average person—took thousands of years. By Descartes's time, however, a certain amount of scientific momentum appeared to be building. While Aristotle, Galen, al-Jahrawi, and other trailblazers couldn't sit on a panel together and discuss their findings, at least their work was being held up for side-by-side examination.

"In the third century BCE, Herophilus demonstrated the existence of sensory and motor nerves, and Erasistratus demonstrated that the brain influenced motor activity. One-half of a millennium later, Galen demonstrated that sectioning the spinal cord caused sensory and motor deficits. Within the spirit of scientific enquiry that resurfaced in the renaissance, anatomical studies by Vesalius published in 1543 reiterated and confirmed Galen's findings. In relation to this, Galen had postulated that three conditions be met for perception: 1) an organ must be able to receive the stimulus, 2) there must be a connection from the organ to the brain, and 3) a processing center that converts the sensation to a conscious perception must exist. Descartes contributed to Galen's model by postulating that a gate existed between the brain and the tubular structures (the connections), which was opened by a sensory cue. A sensory cue would 'tug' on the tube, which would then open a gate between the tube and the brain. The opening of this gate would then allow 'animal spirits' to flow through these tubes and within the muscles to move them. Although this sensory system was not specific to pain, La Forge's drawing (based on Descartes' concept and La Forge's understanding of contemporaneous anatomy) of a foot near a flame is one of the most famous figures in neuroscience."[24]

The next landmark researcher, Charles Bell (1774–1842), had more than a theoretical interest in pain. For much of his adult life, he was bothered by intense discomfort from angina, a condition that eventually killed him. During his career, however, he did much to advance the cause of pain scholarship.

Bell was a Scottish anatomist and surgeon "whose original ideas on the nervous system have been equated with those of William Harvey on the circulation. He suggested that the anterior and posterior nerve roots have different functions, and based on their connectivity he showed that different parts of the brain have different functions. He noted that individual peripheral nerves actually contain nerve fibers with different functions, that nerves conduct only in one direction, that sense organs are specialized to receive only one form of sensory stimulus, and that there is a sixth (muscle) sense."[25]

After Bell, the solution to the pain riddle seemed just over the horizon. Still, it was nearly another century before a later British researcher, Charles Scott Sherrington—the man who discovered and named nociceptors—guided the seemingly endless quest toward the finish line.

Based on Sherrington's research, nociceptors were highly sensitive neurons, activated whenever something occurred in an area of the body that threatened tissue damage. Using the super-highway that is the spinal column, these sentries flash messages to the brain, which responds by sending pain to the affected area as a warning, all of this occurring in an instant. Nor was that all Sherrington uncovered, even though he had no access to many of the diagnostic devices physicians employ today. He began investigating the presence and influence of nociceptors as early as 1906, when the automobile was still a recent invention and the airplane a work in progress. "To many, Charles Scott Sherrington is best known for providing us with the term 'synapse,' a word we still use to describe the junction where two neurons communicate. While Sherrington's work to understand synapses and neural communication was important, however, his studies of reflexes, proprioception, spinal nerves, muscle action, and movement were much more expansive and probably even more influential." Regardless, his observations concerning synapses are representative of the meticulous care with which he investigated and made deductions about the nervous system and its function. His writings about the synapse came at a time when Santiago Ramon y Cajal was beginning to convince the scientific community that the brain consists of separate nerve cells (which became known as neurons in 1891) rather than a continuous "net" of uninterrupted nerves. One thing missing from this theory was an understanding of how neurons might communicate with one another.[26] A dogged and diligent researcher with a distinguished academic record, Sherrington first became interested in nerve reaction and cooperation by observing the mundane reflex test still used by doctors today, tapping the side of the knee to invoke a slight jerk from the leg. How and from where, Sherrington wondered, was that reflex initiated?

So revolutionary was Sherrington's research that he shared the Nobel Prize for Physiology or Medicine with Edgar Adrian in 1932. In his introduction of Sherrington at the Nobel Prize awards ceremonies on December 10, 1932, Professor G. Liljestrand of the Royal Caroline Institute said: "Of fundamental importance for our knowledge of the workings of the nervous system was the discovery that an external influence, a so-called stimulus, can, without the cooperation of the will, call forth a definite response, such as the contraction of certain muscles. A well-known example is presented by the involuntary blinking at a loud and unexpected noise. The external influence is, so to speak, thrown back or reflected, from which the phenomenon received the name 'reflex.' For every one of our movements, even under the influence of the will, for numerous processes in the interior of the body, and in all probability also for mental life itself in its various forms, the reflexes play a highly important role. As a rule, they are provoked by cooperation between groups of afferent, internuncial, and efferent neurons. Sir Charles Sherrington has made extraordinary contributions to our knowledge of the reflex phenomena. In exact experiments employing quantitative methods, he has investigated numerous reflexes, and also single neurons, with the object of establishing general laws for the origin and cooperation of the reflexes in the organism."[27]

A man of broad interests, Sherrington was also a published poet and competed on his school's rugby and rowing teams. By the time he finished his academic career, he had been awarded an honorary doctorate at twenty-eight universities. Although Sherrington's findings may have solved a riddle, they didn't spell the end of research into the origins of pain. As we will see, the phenomenon known as chronic pain continues to bewilder scientists and frustrate physicians, and that's just one of the remaining puzzles. If nothing else, though, pain's public image had been vastly improved—from a stern instrument of punishment to a helpful ally.

This had to be progress.

Chapter 2

The Body's CEO

One of the best and most concise definitions of the human brain I've ever read was in a paper from the American Association of Neurological Surgeons. According to the AANS, the brain "gives meaning to things that happen in the world surrounding us."[1]

Of course.

True, even plants can react to various stimuli—the cooler temperatures in early autumn that cause them to lose their leaves and the warmth of spring that entices those leaves to grow back. And although it's hard to plumb the minds of dogs, horses, and other higher animals, it's obvious that some of their actions have complex thoughts behind them ("If I stop barking, my human will give me a treat."). Yet all of this pales in comparison with the human brain, which takes interpretation of the outside world to a higher level.

"The brain creates our thoughts," explains inventor and futurist Ray Kurzweil, "and our thoughts create our brains."[2] Our thoughts leap past mere observations ("cold," "hot," "large"), into manifestations of judgment ("pretty," "off-key," "ironic").[3] "Fundamentally, the brain is an information processor," Kurzweil says.[4] This means it processes, not simply reacts. It asks not only "what is that?" but "what does that mean?"

Through our discoveries about the origins of pain, we have learned that our brains sometimes operate beyond our consciousness. They are running the show, whether we realize it or not, proactively sending out pain signals like a fire siren alerting a neighborhood.

We don't ask for this—nor, in most instances, desire it—but the brain knows we need it.

Nevertheless, it took quite a few centuries before this most important bodily organ began to get the respect it deserved.

According to a 2012 article in the *Journal of Neurosurgery*, "When ancient Egyptians practiced mummification, the brain was usually liquefied and pulled out from the cranium through the nose using a hook-like tool—a method known as excerebration. You do this by making a hole in the back of the neck and withdrawing it through the foramen, which is the opening at the bottom of the skull where the spinal cord exits the cranium."[5]

The Greek writer Herodotus described this practice around the fifth century BC: "Since the brain was not perceived as important as the heart, it was deemed useless for the afterlife, and so it was disposed of."[6]

Perhaps this explains the mindless behavior of the mummies in 1950s horror movies.

But think about it: liver, kidney, and heart transplants have become almost routine today, but no one has ever succeeded in transplanting a brain. That in itself speaks to the critical role it plays not only in the mere maintenance of life, but also in the element we call "existence."

The physiological challenges involved in a brain transplant would be daunting enough. All the myriad nerves that serve to inform the brain would have to be reinstalled, not to mention the array of blood vessels needed to feed it.

Beyond that, the brain is what makes each person who they are. Tucked away inside our skulls are knowledge, emotions, and memories that belong to no one else. In his dense, fact-filled, and fascinating book *The Body: A Guide for Occupants*, Bill Bryson offers this apparent contradiction: "The great paradox of the brain is that everything you know about the world is provided to you by an organ that itself has never seen that world. The brain exists in silence and darkness, like a dungeoned prisoner. It has no pain receptors, literally no feelings. It has never felt warm sunshine or a soft breeze. To your brain, the world is just a stream of electrical impulses, like taps of Morse code.

"And out of this bare and neutral information it creates for you—quite literally creates—a vibrant, three-dimensional, sensually engaging universe. Your brain is you. Everything else is just plumbing and scaffolding."[7]

As neuroscientist Antonio Damasio reflected in a December 19, 2011, TED talk: "The wonder is about the fact that we all woke up this morning and we had with it the amazing return of our conscious mind. We recovered minds with a complete sense of self and a complete sense of our own existence, yet we hardly ever pause to consider this wonder. We should, in fact, because without having this possibility of conscious minds, we would have no knowledge whatsoever about our humanity; we would have no knowledge whatsoever about the world. We would have no pains, but also no joys. We would have no access to love or to the ability to create."[8]

So if the brain of someone like Stephen Hawking could be transplanted into a person with mental retardation, would that person automatically become a brilliant physicist? Or would the transplantation process unplug all that, leaving the brain's new owner with nothing but a blank slate? Such intriguing speculation suddenly became an issue in the scientific and medical communities in 2017, when Italian neurosurgeon Sergio Cavanero announced that he had performed the world's first successful "human head transplant" using two human cadavers in China.

Described in the London *Telegraph* as "controversial" (a masterpiece of an understatement), the operation was said to have taken around eighteen hours. Moreover, Cavanero reported that he expected to be able to carry out the same procedure on a living person "imminently."[9] The two individuals involved in the transplant had no comment, being dead. But Arthur Caplan, a professor of bioethics at New York University's Langone Medical Center and head of the US organ distribution center, did, calling Cavanero's announcement "the continuation of a despicable fraud."

"We have a face transplant program here [at NYU]," Caplan continued. "It is very difficult to just transplant the face. It requires massive doses of immunosuppressants. The head would be an even bigger problem, requiring even bigger doses. It would probably kill you in a few years from rejection or infection. If he [Cavanero] knew how to get the spinal cord to repair, to reconnect, he should be doing it on people with spinal cord injuries. There are millions of such people around the world. They want to walk; they want to control their bodies, their bowels. There is no reason not to go there and show what you can do."[10] Meanwhile, four years later, "imminently" has yet to arrive for Sergio Cavanero. This controversy was a reminder of the human brain's incredible properties. Not only can it retain and store vast amounts of information, but it can allow us to "remember" such things as tastes, sounds, and smells. It's what puts us to sleep at night and wakes us up in the morning. It allows us to walk and balance (as anyone who has suffered a debilitating stroke knows quite well). If you think "I want that brownie," the brain is what directs your hand to reach out and grab it, just as it tells you to quickly remove that hand if it brushes against the surface of a hot stove. Truth be told, it isn't the supposedly romantic heart that causes us to feel sexual attraction for another person, but the brain—and should that attraction lead to something more, the brain usually initiates all the necessary physical reactions to make that possible. We tend to take a lot of this for granted, but the rise in Alzheimer's disease (perhaps because we are now living longer) has put the memory function of the brain under a microscope. It's why those of us past the midpoint of life start to worry if we can't immediately come up with the name of the lead actor in that movie we just watched.

Memory is not only remarkable but occasionally capricious. Just as an example, try telling yourself, "I'm going to focus my thoughts on remembering something random from my past," and clear your mind. You might be amazed by what pops up, perhaps a seemingly insignificant slice of past life you thought was long forgotten. Why had your brain saved that in such detail, when you may no longer have a firm and complete recollection of your college graduation or wedding day?

"Memory storage is idiosyncratic and strangely disjointed. The mind breaks each memory down into its convenient parts—names, faces, locations, contexts, how a thing feels to the touch, even whether it is living or dead—and sends the parts to different places, then calls them back and reassembles them when the whole is needed again. A single fleeting thought or recollection can fire up a million or more neurons scattered across the brain. Moreover, these fragments of memory move around over time, migrating from one part of the cortex to another. It's no wonder that we get details muddled."[11]

On the other hand, why do some people display what is called a "photographic memory"?

In the movie *Rain Man*, Tom Cruise's Charlie Babbitt character watches with amused disbelief as his brother Raymond—an autistic savant—memorizes a hotel phone book. But that night when they're out to dinner, Raymond sees the name tag on their waitress's uniform and startles her by rattling off her phone number.[12]

Waitresses don't wear name tags anymore, phone books are disappearing, and not everyone is a savant, but you get the idea. In fact, the movie was based on a real person (and, thus, a real brain).

In 2017, the *Guardian* reported on Jill Price, one of the first individuals to be diagnosed with highly superior autobiographical memory (HSAM):

"If you ask Jill Price to remember any day of her life, she can come up with an answer in a heartbeat.

"What was she doing on 29 August 1980?

"'It was a Friday, I went to Palm Springs with my friends, twins, Nina and Michelle, and their family for Labour Day weekend,' she says. 'And before we went to Palm Springs, we went to get them bikini waxes. They were screaming through the whole thing.' Price was 14 years and eight months old.

"What about the third time she drove a car?

"'The third time I drove a car was January 10, 1981. Saturday. Teen Auto. That's where we used to get our driving lessons from.' She was 15 years and two weeks old.

"The first time she heard the Rick Springfield song 'Jessie's Girl'?

"'March 7, 1981.' She was driving in a car with her mother, who was yelling at her. She was 16 years and two months old."[13]

Like most rare gifts, however, HSAM has its downside.

"Whenever I see a date flash on the television (or anywhere else for that matter)," Price said, "I automatically go back to that day and remember where I was, what I was doing, what day it fell on and on and on and on and on. It is non-stop, uncontrollable, and totally exhausting. . . . Most have called it a gift but I call it a burden. I run my entire life through my head every day and it drives me crazy!!!"[14]

Needless to say, Price has attracted more than her share of research interest.

Writes Yvaine Ye in *New Scientist*: "Highly Superior Autobiographical Memory (HSAM) is a condition that has been identified in fewer than 100 people worldwide. . . . Their memories are exceptional, but certainly not as perfect as actual photographs. Doctors have yet to understand what exactly happens in the brains of individuals with HSAM, and tests reveal there is no particular ability that appears to underpin the condition. Results from one study suggest that people with HSAM are no better at acquiring memories—they are not superior learners—but are simply better at retaining memories."[15]

If anything is specifically responsible for Price's mixed blessing, it is her hippocampus, a region of the brain where long-term memories go to die (or, in her case, apparently, live forever).

According to the Encyclopedia Britannica, "The name *hippocampus* is derived from the Greek *hippokampus* (*hippos*, meaning 'horse,' and *kampos*, meaning 'sea monster'), since the structure's shape resembles that of a seahorse. The hippocampus, which is located in the inner (medial) region of the temporal lobe, forms part of the limbic system, which is particularly important in regulating emotional responses. The hippocampus is thought to be principally involved in storing long-term memories and in making those memories resistant to forgetting, though this is a matter of debate. It is also thought to play an important role in spatial processing and navigation."[16] Where the hippocampus sometimes turns against us is in the area of chronic pain. The brain in such cases might be reacting not to the stimuli at hand, but to memories of a past painful experience. "The problem is that for all the wonderful things our brain does, it has a hard time forgetting pain. In fact, research shows that any pain lasting more than a few minutes leaves a trace in the nervous system."[17]

In 2012, a team of researchers at McGill University in Montreal looked into that scenario. "They knew that because of the traces left behind, people with chronic pain often develop a hypersensitivity to more pain or even touch. They also knew that a protein enzyme called PKM-zeta plays a critical role in building and maintaining memory by strengthening connections between neurons. So they set out to see if the PKM-zeta was responsible for pain memories and if they could erase them by blocking its activity at the neuron level.

"This is where lab rats enter the picture, but in this case, so do chili peppers . . . or more accurately capsaicin, the compound that makes them burn. The scientists applied capsaicin to the rats' back paws, giving them a chemical hotfoot. And that's when they discovered that PKM-zeta built up in the animals' central nervous systems. Then, after applying to the rats' spinal cords a chemical known as ZIP—which has been shown to stop the brain from holding on to memories—they found that the paws were no longer sensitive. The pain memory had gone away.

"Or as Terence Coderre, the neuroscientist who headed the research put it: 'We were basically able to erase it after the fact.'

"We're still a long way off from ZIP becoming a pain treatment. Obviously, a lot of hurdles would need to be cleared, such as how to wipe out only pain memories without also losing recollections of your first kiss or the last time you got a great parking spot. But Coderre and his team have identified a target. And we're a little bit closer to pain that truly is fleeting."[18]

Just as *Rain Man* offered a fictional outtake on HSAM, *Eternal Sunshine of the Spotless Mind* dealt with a future treatment capable of erasing unpleasant memories forever. In that film, Jim Carrey and Kate Winslet attempt to exorcise all recollections of their relationship gone wrong with unexpected results. Most of us now give the brain the credit it's due. Nevertheless, it wasn't until the late twentieth century that the average person had a frame of reference for it.

"Oh, I see. It's like a computer." Indeed, it is. Both store information that can later be retrieved. Both serve as the nerve center (literally, in the case of the brain) of a complicated communication system. Thanks to the development of highly sophisticated technology, we now have machines and systems that store the information that is then translated into artificial intelligence (AI) and is used to execute complex activities including driving an otherwise driverless car. Yet both brain and computer are vulnerable to glitches called—in both cases—"viruses." Perhaps another

analogy is also appropriate. With its diverse but efficient array of parts, all working toward the same goal, the human body could be compared to a corporate entity. If that's the case, the brain is the CEO. Although it weighs only around three pounds and is composed mostly of soft tissue, the brain is unquestionably the body part in charge. As with many corporate leaders, its office is located on the top floor of company headquarters.

Nevertheless, that comparison also goes only so far. To begin with, not even the most work-obsessed CEOs are able to remain on the job around the clock, every second of the day. The brain, however, never sleeps. And it never slows down. Moved along by its billions of neurons, the speed of information rocketing around in the brain has been estimated at roughly 1,250 miles (402 kilometers) an hour.[19] The sheer volume of this information is equally impressive. It's estimated that your brain can store 2.5 million gigabytes.[20] Moving to the United States from Kashmir required me to read, write, and speak in a new language. The most challenging part of this linguistic makeover for most people involves conversation.

That's because people don't speak in words, in the practical sense, but in sentences or phrases. If someone remarks, "It's a nice day today, isn't it?" you're not hearing "It's. A. Nice. Day. Today. Isn't. It?" but something more akin to "Itsanicedaytodayisntit?" That's why those speaking a language unfamiliar to us seem to be talking so fast. Even if you've learned to translate all the individual words, trying to interpret them one at a time leaves you hopelessly behind in the conversation. Eventually, however, most brains can handle it. The more gifted linguists can juggle a dozen different languages, moving back and forth among them with ease. And it isn't just words and names and phone numbers.

"Our visual brains have vastly more storage than our linguistic brains. Prove this to yourself by conjuring up an image of the inside of your house or apartment. Notice that as you mentally walk through the residence, you can remember where every room is, and even 'see' details of what is in each room, such as beds, furniture, and art hanging on walls. That's a huge amount of information you've just recalled."[21] Still, although a vast memory function is a nice accessory for brain owners to enjoy, it isn't absolutely essential. Someone with advanced dementia may have no memory at all, especially in the short term, yet the brain keeps countless other functions operating. It is, after all, the CEO.

Like most CEOs, however, the brain's services don't come cheaply. It appropriates 50 percent of the blood pumped by the heart and uses far more than its share of oxygen. At the same time, as Bill Bryson pointed out, it performs its myriad functions out of sight, while the heart calls attention to itself with its steady, life-affirming beat and the kidneys and the stomach routinely remind us of their functions every day. No wonder the early explorers of the human body were confused. They were actually using their brains to prove that the brain was unnecessary. The earliest known reference to the brain comes from what became known as the Edwin Smith Surgical Papyrus, which was found in Egypt and dates back to the seventh century BC. The author was a battlefield surgeon who noted some of the symptoms resulting from head injuries and even described the appearance of an exposed brain.[22]

The ancient Greeks pushed that rudimentary knowledge ahead a bit with the Pythagorean Almaeon of Croton (fifth century BC), credited as the first writings to describe the brain as the location of the mind. "All the senses are connected in some way with the brain," it contended, "consequently, they are incapable of action if the brain is disturbed."[23]

Any physician who deals with head injuries can recall instances in which damage to one part of the brain affected only a few of its functions, much as one blown fuse in your fuse box might shut down only your dishwasher or the lights in your bedroom but not the whole electrical system.

Here's another example: "You walk into your child's daycare center. As usual, there are a dozen kids there waiting to get picked up, but this time, the children's faces look weirdly similar, and you can't figure out which child is yours. Do you need new glasses? Are you losing your mind? You run through a quick mental checklist. No, you seem to be thinking clearly, and your vision is perfectly sharp. And everything looks normal except the children's faces. You can see the faces, but they don't look distinctive, and none of them looks familiar, and it's only by spotting an orange hair ribbon that you find your daughter.

"This sudden loss of the ability to recognize faces actually happens to people. It's called prosopagnosia, and it results from damage to a particular part of the brain. The striking thing about it is that only face recognition is impaired; everything else is just fine."[24]

We now see the brain not as a single unit, but a complicated partnership of different elements, more like a condominium than a house.

The National Institutes of Health published a report called "Brain Basics" that is worth quoting at some length:

"The brain is like a committee of experts. All the parts of the brain work together, but each part has its own special properties. The brain can be divided into three basic units: the forebrain, the midbrain and the hindbrain. The hindbrain includes the upper part of the spinal cord, the brain stem, and a wrinkled ball of tissue called the cerebellum. The hindbrain controls the body's vital functions such as respiration and heart rate. The cerebellum coordinates movement and is involved in learned rote movements. When you play the piano or hit a tennis ball you are activating the cerebellum.

"The uppermost part of the brain stem is the midbrain, which controls some reflex actions and is part of the circuit involved in the control of eye movements and other voluntary movements. The forebrain is the largest and most highly developed part of the human brain: it consists primarily of the cerebrum and the structures hidden beneath it.

"When people see pictures of the brain it is usually the cerebrum that they notice. The cerebrum sits at the topmost part of the brain and is the source of intellectual activities. It holds your memories, allows you to plan, and enables you to imagine and think. It allows you to recognize friends, read books, and play games.

"The cerebrum is split into two halves (hemispheres) by a deep fissure. Despite the split, the two cerebral hemispheres communicate with each other through a thick tract of nerve fibers that lies at the base of this fissure. Although the two hemispheres seem to be mirror images of each other, they are different. For instance, the ability to form words seems to lie primarily in the left hemisphere, while the right hemisphere seems to control many abstract reasoning skills.

"For some as-yet-unknown reason, nearly all of the signals from the brain to the body and vice-versa cross over on their way to and from the brain. This means that the right cerebral hemisphere primarily controls the left side of the body and the left hemisphere primarily controls the right side. When one side of the brain is damaged, the opposite side of the body is affected. For example, a stroke in the right hemisphere of the brain can leave the left arm and leg paralyzed."[25]

Got that? Good, because it gets a lot more complicated. Those two hemispheres are themselves subdivided into many specific working areas continually bombarded with nerve impulses and often at the mercy of naturally occurring hormones. This is the chaotic world where pain is born, lives, and dies.

Fortunately, it is also where the brain asserts itself.

Some aspects of human existence have changed significantly over the centuries, with a corresponding impact on our bodies' relationship with pain.

Along with sending pain to warn us of potential injury or illness, the brain also presides over the well-known "fight or flight" impulse. Though that may have served a more direct and urgent function with early man ("Don't look now, but there's a saber-toothed tiger right outside your cave!"), much of what stresses us out today aren't things that can be dealt with by a well-aimed spear.

"If you live under chronic stress, working in a stressful job, or remain stuck in a toxic relationship while focusing on the negative aspects of your life, your sensation of pain after a physical injury can be intensified and your body's healing process disrupted. Moving beyond pain becomes difficult.

"The pain response is your body's way to protect you. Your brain literally thinks that an injury or emotional stress is placing you at risk, so it creates pain to force you to withdraw from the situation and seek healing.

"Chronic stress puts your nervous system in a highly reactive state. Your brain sends danger signals to your body, resulting in overstimulation of the sympathetic nervous system, which in turn leads to the excess production of your stress hormone cortisol. Among its functions, cortisol raises blood sugar and suppresses the immune system.

"Increased cortisol levels release chemicals that cause inflammation, aiding healing and recovery. Cortisol should do its job and simmer down. When this hormone sticks around beyond its prime, it inhibits healing and contributes to poor sleep, anxiety, and depression, all of which exacerbate pain and lead to weight gain."[26]

In 2015, a University of Colorado study took a new direction, positing that the brain might have more power over pain than was previously believed.

"Does cognitive self-regulation influence pain experience by affecting the primary representations of painful (nociceptive) stimuli in the brain? Or does it regulate reported pain via a neural pathway that is distinct from the one that mediates nociceptive pain? The present study demonstrates that nociceptive and cognitive manipulations of pain influence two distinct, separable neural systems, which operate together to construct the pain experience. The neurologic pain signature (NPS) mediates the effects of noxious input, whereas a fronto-striatal pathway connecting the nucleus accumbens and ventromedial prefrontal cortex mediates the effects of cognitive self-regulation of pain. These findings help move the field beyond the 'one system' view of pain as a primarily nociceptive process, and provide a foundation for new approaches to multidimensional pain assessment and treatment."[27]

Cutting through the medical jargon, this means it may not be necessary to deactivate or otherwise interfere with the message delivered by the nociceptors. Rather, the Colorado study says the brain seems to be capable of creating a new "neural pathway" to allow the passage of messages that "self-regulate" pain. Rather than turning off the nociceptors, this would, in effect, provide the brain with a second opinion.

Three years before the Colorado study, psychiatric counselor Glynis Sherwood published some of what she had learned from working with marathon runners and other extreme athletes.

"Central to the success of marathon runners is the ability to cultivate an affirmative mindset, specifically a positive attitude and optimistic expectations from their efforts. They focus on visualizing themselves achieving their goals and mentally distancing themselves from the discomfort and pain that are part of any long-distance race. These are psychological skills that are key to any long-term approach to breaking free from long-standing emotional distress."[28]

One of the amazing things about the brain is that it can be trained to react to repetitive information. We don't need to relearn how to tie our shoes each time it becomes necessary or how to walk or how to start and drive a car. Otherwise, our average days would be infinitely longer and more complicated.

It could be said that the competitive runners whom Glynis Sherwood counsels deal with chronic pain of their own creation, inviting and accepting the significant discomfort they know they will feel in the heart of every race. Yet because this pain is recurring, they have been able to create an alternative message to the brain that says: "I know this hurts, but it's OK."

Of course, the ultramarathoner realizes that his or her pain will dissipate soon after the event is over, an optimistic insight not shared by those grappling with chronic pain of unknown origin.

Still, Sherwood's findings hold out hope that if a nonathletic chronic pain sufferer can somehow find a way to deflect the emotional baggage that comes with that, they too can send a positive alternative message "upstairs."

Thus, the brain affects our relationship with pain on two levels, the physical and the emotional.

"Pain is hard-wired as a social alarm of a threat, which is then selected over other competing demands and triggers behaviors that interfere with normal life functioning. Each individual's experience of pain and its expression is a product of the sensory experience, the person's personal background, the interpersonal context, and the meaning it has for the individual."[29]

In 1965, Ronald Melzack and Patrick Wall offered what they called the gate control theory, which described "how non-painful sensations can override and reduce painful sensations. A painful, nociceptive stimulus stimulates primary afferent fibers and travels to the brain via transmission cells. Increasing activity of the transmission cells results in increased perceived pain. Conversely, decreasing activity of transmission cells reduces perceived pain. In the gate control theory, a closed 'gate' describes when input to transmission cells is blocked, therefore reducing the sensation of pain. An open 'gate' describes when input to transmission cells is permitted, therefore allowing the sensation of pain."[30]

Wrote Melzack: "Pain is a personal, subjective experience influenced by cultural learning, the meaning of the situation, attention, and other psychological variables. Pain processes do not begin with the stimulation of receptors. Rather, injury or disease produces neural signals that enter an active nervous system that (in the adult organism) is the substrate of past experience, culture, and a host of other environmental and personal factors. These brain processes actively participate in the selection, abstraction, and synthesis of information from the total sensory input. Pain is not simply the end product of a linear sensory transmission system; it is a dynamic process that involves continuous interactions among complex ascending and descending systems."[31]

This unlocked a new frontier for pain researchers. For one thing, it helped to explain why the pain from an abrasion or bump is lessened by rubbing the area. The impulse that causes us to do that is tied to a group of neurons Melzack and Wall called non-nociceptors, which compete with the pain signal for the brain's attention. Despite some critical responses, the gate theory has remained an important part of pain theory for more than fifty years.

When Ronald Melzack died in 2020 at the age of ninety, the *Montreal Gazette* included a quote from Jeffrey Mogil, a behavioral neuroscientist at McGill University, where Melzack taught: "I think a case can be made that Ron Melzack had the greatest career of any person to ever be in the field of pain research."[32]

Melzack was inducted into the Canadian Medical Hall of Fame in 2009. The citation accompanying that honor declared: "It's been said that Dr. Ron Melzack has done for pain research and pain management what Einstein did for physics."

The *Montreal Gazette* article also reprised a quote from Melzack that neatly encapsulated his long career: "'No one should have to feel pain,' Melzack said."[33]

Chapter 3

Pain in Our Culture

The title of a 1980s tune by the alternative rock band REM may have said it best: "Everybody Hurts."

Everybody, that is, except for that tiny subset of individuals who are unable to feel any pain at all—a mixed blessing, as we shall see. For the rest of us, pain is one of the three universal human experiences, tucked in between birth and death.

"The aim of the wise," observed the Greek philosopher Aristotle, "is not to secure pleasure, but to avoid pain."

Unfortunately, pain often refuses to be avoided, even by the wise. And although it sometimes retreats almost as quickly as it arrives, severe or long-lasting pain must be confronted on many levels. That has prompted a massive army of opposition to rise up over time, to the point that pain relief in the United States has become one of the country's major industries.

Soldiers in that army include the 6.2 million people employed in American hospitals, the million-plus practicing physicians, and the more than three hundred thousand pharmacists, not to mention the ultra-lucrative pharmaceutical industry.

Indeed, the universal aspect of pain represents a potential gold mine for those who make it their business. If everybody hurts, then everybody will, at one time or another, require some form of pain medication. Thus, "Big Pharma" in the United States took in upward of $490 billion in 2019, more than 4 percent of the US gross domestic product.

"Obviously, pain relievers are for every demographic," noted T. S. Kelly on the *Drug Store News* website. "The split between female-centric programming and rough-and-tumble sports fare makes it clear that no matter what you're watching from the couch, OTC drug makers recognize that aches and pains are an everyday problem for every one of us."[1]

As a result, TV and online ads for pain relievers are now almost as ubiquitous as the pain itself. The current array of products ranges from pills to liquids to numbing sprays to hypnotism, most claiming to have an edge in both pain management and the swiftness with which that relief can arrive. Some of these ads come from reliable sources; others veer off into "snake oil" territory.

Not even the opioid epidemic has significantly diminished this windfall. Many companies simply pivoted in a different direction—the US topical pain relief market was valued at $26 million in 2019 and is projected to reach $57 million by 2027.[2]

To quote one industry press release: "Topical pain relief medications are applied directly to the epidermal layer of the skin at the area of inflammation or pain. The aim is to offer a slow and gradual release of pain relief medication into the bloodstream by keeping the blood levels relatively constant for a certain period of time. These medications offer lesser side effects as compared to oral medications."[3]

In other words, they aren't OxyContin. But even Purdue Pharma, run by the Sackler family and hit with enormous fines by the federal government for misrepresenting the addictive power of certain opioids, apparently replicated its previous strategy in offshore markets such as China.

According to the Associated Press, "Documents and interviews indicate that representatives from the Sacklers' Chinese affiliate, Mundipharma, tell doctors that time-release painkillers like OxyContin are less addictive than other opioids—the same pitch Purdue admitted was false in U.S. court more than a decade ago. Mundipharma has pushed ever larger doses of opioids, even as it became clear that higher doses present higher risks, and represented the drug as safe for chronic pain, according to the interviews and documents."[4]

Whatever the anti-pain product, though, marketers know that modern humans are an impatient breed, geared toward instant gratification. We have fast food for hunger, online dating for loneliness, and multiple internet search engines that can answer virtually any of our questions in seconds. Why should help for pain be any different?

"Order now," promises one internet come-on, "and the pain is gone."

A company in the United Kingdom used the slogan "Don't let pain hold you back" and encouraged consumers to "stop pain from ruining your plans."

Of course, if that pain comes from a broken leg or colon cancer, your plans are probably ruined, anyway—and the discomfort you feel is a warning, not an inconvenience. Trying to isolate pain from its cause is not only ill-advised but usually fruitless. Medication may make the underlying condition hurt less—or, temporarily, not at all—but it is a short-term fix when long-term treatment is required.

Thus, the sad story that doctors sometimes hear from patients eventually diagnosed with an "end-stage" medical condition: "I just kept taking painkillers until it got too bad to stand it. That's when I came to see you."

If I don't hurt, I must be OK. That's what the TV ads—and our own wishful thinking—tell us. But this approach is akin to trying to turn off an annoying fire alarm without realizing that your house is ablaze.

Not so long ago, professional athletes were often injected with novocaine or some other painkiller after being injured then sent back on the field to do battle. It was the pain, not the injury, that was perceived as the problem.

Without question, our attitudes toward pain are often complicated. Whereas some cultures demand immediate relief, others, like those numbed athletes, pride themselves on their stoicism. Yet struggling to disregard pain, no matter how noble or courageous it seems, can be as misguided as overmedicating it.

There is also some truth to the old saying "misery loves company." That's why the internet is full of online sites welcoming pain sufferers.

"One of the things a support group can do is to help people in pain know they are not alone," says Richard A. Lawhern, who has moderated online support groups for more than twenty-five years. "They can offer support to people . . . whose caregivers may be getting burned out."[5]

Certainly, complaints about the painful effects of a broken body can, over time, become a broken record. No matter how sympathetic a family member or friend may be, continued exposure to the same lament can be both exhausting and numbing. Physicians might offer empathy, but their busy schedules necessarily minimize "face time" with patients. Moreover, they have seen it all before—and in the case of specialists, many times over.

The arrival of the COVID-19 pandemic in 2020 only heightened the sense of isolation for many chronic pain sufferers. In many cases, enforced quarantines and social distancing wiped out what little social interaction had been previously available. Even after some of these restrictions were eased or removed, chronic pain patients often found their place in the health-care system profoundly changed.[6]

"Although pain treatment has been described as a fundamental human right, the Coronavirus disease 2019 (COVID-19) pandemic forced healthcare systems worldwide to redistribute healthcare resources toward intensive care units and other COVID-19 dedicated sites. As most chronic pain services were subsequently deemed non-urgent, all outpatient and elective interventional procedures have been reduced or interrupted during the COVID-19 pandemic in order to reduce the risk of viral spread."[7]

To someone suffering a major flare-up from back pain or fibromyalgia, "nonurgent" is a matter of opinion, and all this made internet support groups for those marooned by pain even more critical. Some groups are practical in nature, sharing information about various medications, treatments, and exercise routines; others exist merely as sounding boards and cries for help.

These posts on the website *Daily Strength* are typical:

> "It's been about 9 years since the wreck. Narcotics are working, but I know there will come a time when they won't."
>
> "Horrible pain this morning made me off-mood. Could use some kind words today."
>
> "Has anyone here had microvascular decompression surgery?"[8]

Many resources are designed to serve very specific niches and communities. The advantage here is that a question about migraine headaches or chronic backaches will be met with knowing nods rather than confusion.

Social media sites such as Facebook or Instagram offer additional avenues for venting, although the reaction from a broad general audience is often mixed.

"It's kind of weird," notes one Facebook user, "when you post something about being in pain and all these people respond with 'like.'"

Meanwhile, some of the online information purporting to be educational turns out to be cleverly disguised ads for specific drugs or doctors' practices. Often, these lure readers with inflammatory headlines designed to create fear and cite the "approval" of government agencies and medical societies.

Other scams arrive by telephone, such as one robocall reported by the *Morning Call* newspaper of Allentown, Pennsylvania. That pitch began: "In an attempt to stop the growing use and abuse of prescription narcotic pain pills, America's national health-care providers are now authorized to provide a new experimental pain relief compound to anyone suffering from physical pain or discomfort.

"This compound cream is directly applied to the pain-related areas," it continued. "It is non-narcotic and extremely effective. This new pain relief compound is provided to you by your insurance carrier with no out-of-pocket expense to you, and your pain relief cream can be shipped immediately."[9]

The vague tone of this "phishing" effort is a tip-off. Since when do "America's national health-care providers" unite en masse behind a single product? And if this miracle cream is "experimental," why has it been approved at all?

As for the promise of "no out-of-pocket expense to you," it seems obvious that the consumer is expected to pay first, then be reimbursed by his or her insurance carrier. Except that those who were fooled by this scam soon discovered that the existence of this product was news to their insurers.

Nevertheless, those who push pain relief products, legitimate or not, have a unique marketing advantage. Pain is something that can't be ignored.

It is also highly personal.

All human hearts, livers, and kidneys are essentially the same—hence, the donor transplant question on most state driver's licenses. But there are no brain transplants, as we've mentioned, because every human brain is unique. Since most pain is a brain function, the same injury or affliction can affect persons A and B in different forms and to varying degrees.

If you've been to a hospital or doctor's office in the last few years, you've no doubt seen one of those "pain indicator" charts attached to a wall. The original version was known as the visual analog scale (VAS), which required the patient to point to a spot along a line, the left end of the scale indicating no pain and the right the worst pain imaginable. That has since been largely replaced by the Wong-Baker FACES scale, in which a child is shown six faces ranking the degree of pain from one to ten. Faces one and two are smiling almost giddily, and nine and ten are twisted into grimaces.

Though they may be useful at times, the problem with these charts is that they foster the illusion that pain can be given an objective "score." Realistically, most of us simply guess at a number, reconfirming the personal aspect of pain. What might be a "five" for one individual could be a "seven" or a "three" for others, and that assessment could depend on the day. Thus, the physical condition causing the pain is not as relevant as the way each person's body interprets it.

"Pain is inherently subjective and could be considered an emotional response to a personal experience. In fact, emotional suffering is an important and perhaps underappreciated aspect of persistent pain. At the same time, pain is ubiquitous and familiar to everyone and remains one of the most common reasons that Americans access the health care system."

With chronic pain patients, however, "there is frequently no apparent noxious stimulus. [The degree of pain] is typically measured either based on the patient's self-report or on observation of the patient's behaviors, which may lead to unreliable results."[10]

In an interview on National Public Radio, Dr. John Markman, director of the Translational Pain Research Program at the University of Rochester School of Medicine and Dentistry, expressed concern that the charts might lead doctors to "treat by numbers" and thus keep patients from getting the most effective care.[11]

To find out more about how the numerical pain scale was affecting doctor-patient communication, Markman and his colleagues at the University of Rochester undertook a study they presented to the World Congress on Pain in Boston in 2018.

Patients were asked to rate their pain on a scale of zero to ten and then asked the question, "Is your pain tolerable?"

Surprisingly, three-quarters of the patients who rated their pain between four and seven on the numerical scale, a range that typically calls for higher doses of medications, also described their pain as "tolerable"—a description that normally means no more pain treatment is needed.

This showed the danger of relying only on a number, Markman said. "If you were just treating by the numbers, you might say, 'Well, someone has a pain that is 6 [out of] 10. I feel obligated to do something about that . . . to fix that number just like you might fix their blood pressure or their blood glucose.'"[12]

A device called the dolorimeter became popular in the mid-twentieth century, which used various methods—such as a focused beam of light—to inflict pain and record a subject's reactions to it. Several pharmaceutical companies used it to promote their pain relief products, and *Time* magazine reported in 1945 that Cleveland physician Lorand Julius Bela Gluzek had developed a dolorimeter that actually measured pain in grams.

Gluzek claimed that his device was 97 percent accurate, but subsequent testing cast increasing doubt on the reliability of those and other dolorimeter results. By the mid-1950s, the dolorimeter had fallen out of favor, and subjectivity reigned once again.[13]

This was an issue first addressed by the Greek and Roman physician Galen eighteen hundred years ago, and health-care providers are still waiting for a solution. When clinicians look at a number, Markman says, they may be more likely to overtreat. Likewise, the degree of pain reported by the patient is not always reliable.[14]

In some cases, though, undertreatment becomes more of an issue than overreach. In recent years, hospitals have begun emphasizing a patient's right to be as pain-free as possible. Previously, it often was assumed that the person being treated would bring up his or her pain status voluntarily if it seemed relevant. Otherwise, the attending physician didn't always ask.

"Some doctors are sensitive, some are not," says one former hospital patient. "The doctors that understand pain best are generally ER doctors. Some doctors that specialize in pain management eventually learn about pain if that doctor listens to the patient, not just read the literature.

"You don't know what it feels like to fall off a horse if you only read it in a book. You have to actually fall off the horse. Same with pain and listening to a patient. Each patient reacts uniquely to pain, no matter the source."

Postsurgical pain can pose a special problem. Given the pain-cloaking effects of anesthesia, several hours may have elapsed after the operation before a patient begins to feel serious discomfort. By then, the odds are that the physicians involved in that case and the accompanying surgery are no longer immediately available. Because a physician must sign off on most pain medications, it is sometimes left to nurses or physician assistants to try and contact someone in authority, which can often take hours. Meanwhile, the patient continues to hurt.

In other words, for some doctors, pain might be more of a patient comfort issue than a treatment concern.

The logical solution might be to order pain medication in advance, but sometimes it is difficult to determine what a patient's needs might be. Postoperative pain, like all pain, can depend on the individual and the circumstances, and hospital personnel other than doctors are often left to make decisions about pain relief.

Complicating these choices, even if a physician's order is at hand, is the recent wariness about opioid drugs. A 2016 article in the *Daily Nurse* dealt with that change in medical attitude:

"The patient who watches the clock and requests their pain medication at the top of the hour. The patient who always rates their pain a '10' out of 10. The patient who requests a specific narcotic like Dilaudid. Some nurses might view such behavior as red flags and will label those patients as potential 'drug seekers'—but pain management experts like Bobbie Norris, BSN, RN, CNRN, BC-RN, a pain resource nurse at Johns Hopkins Department of Neurology and Neurosurgery in Maryland, says nurses who jump to those conclusions are often wrong and do a disservice to their patients.

"The patient specifically asking for Dilaudid, for instance, isn't necessarily an addict. In fact, a patient returning to the hospital for his umpteenth surgery most likely *is* an expert on what medications work best for him. 'Just because a patient knows what works for them, that doesn't mean they're drug-seeking,' says Norris.

"Susan McMillan, PhD, ARNP, FAAN, a nursing professor at the University of South Florida who has researched pain in oncology patients, echoes Norris' concerns. 'Nurses today are very concerned about drug-seeking,' she says, recalling a study in which nurses were asked what made them decide if a patient was 'drug-seeking,' as opposed to suffering. Their answers were: 'If their pain was unrelieved, if it's overwhelming, or if they ask too frequently,' says McMillan, though in reality, each of those behaviors is an indicator that a patient's pain is not being well managed.

"Indeed, if Hospital Consumer Assessment of Healthcare Providers and Systems (HCAHPS) scores are an effective measurement, patients are not getting enough pain relief during their hospital stays. According to HCAHPS patient survey data, only 71% of those surveyed said their pain was 'always' well controlled in 2014–2015. Other studies, meanwhile, show that pain is often undertreated in pediatric patients, in older adults in long-term care, and among certain minority populations."[15]

If pain were an intruder, similar to a virus or allergic reaction, its effect on the human body would be more predictable. Since it is triggered by the brain, however, it can be affected by memories, anxieties, or even genetic factors unique to that individual.

Someone who has experienced extreme pain from a particular medical or dental procedure, for example, might find that fearful anticipation escalates the pain the next time. That's what makes long-term torture so effective.

By contrast, someone who has never experienced a procedure is less likely to be tense and on guard.

Most cultures differentiate between "good pain" and "bad pain." The former is usually considered a trade-off in which temporary pain leads to a positive outcome. The pain of childbirth would be one example, the soreness after an exercise workout another. "No pain, no gain" has become a mantra.

Meanwhile, some tribal societies (and college fraternities) use pain as a rite of passage, especially for young men.

It is also a long-held belief among some parents that inflicting pain on disobedient children—normally through spanking—will create a negative memory that might encourage better behavior in the future.

According to University of Colorado instructor Marcia Carteret, "There is a long tradition of stoicism in European American culture; generations of children, especially boys, would be admonished for crying like babies but applauded for keeping a stiff upper lip. In general, people made as little fuss as possible over injuries and illness. Naturally, children socialized in this way will grow up to be 'easy patients' who behave in ways consistent with the values of the western medical system.

"On the other hand, there are cultures where a child's crying immediately elicits the greatest sympathy, concern, and aid. In such cultures, children's health is fretted over constantly—even a sneeze can be seen as an illness. This predisposes children to become more anxious about their health in general, and as adults, they may need greater reassurance from caregivers even in the face of minor symptoms."[16]

Religion can enter the picture, as well. In their research paper "Spirituality and Religion: Pain and Pain Management," Ozen Dedeli and Gulten Kaptan noted: "There is growing recognition that persistent pain is a complex and multi-dimensional experience stemming from the interrelations among biological, psychological, social, and spiritual factors.

"In the Middle Ages, the pain was considered a religious matter. The pain was seen as God's punishment for sins, or as evidence that an individual was possessed by demons. This definition of pain is still embraced by some patients who might tell the health professionals that the suffering is their *cross to bear*. Pain relief may not be appealing to those individuals who believe in this definition of pain. Spiritual counseling thus may be more of a priority than medical management. Many Hindu believers envision pain as a divinely ordained punishment. In Islam, it can be punitive or Allah's will. A popular Buddhist belief is that suffering is the cost of attachment."[17]

Prayer can offer much-needed hope for the devout. The very act of asking a higher power to deliver relief from pain sets up the possibility that help is, indeed, on the way. This alone may make the pain more bearable.

David Weissman, Deborah Gordon, and Shiva Bidar-Sielaff put their research on this subject into an article titled "Cultural Aspects of Pain Management." The article said in part, "Culture is the framework that directs human behavior in a given situation. The meaning and expression of pain are influenced by people's cultural backgrounds. Pain is not just a physiological response to tissue damage but also includes emotions and behavioral responses based on individuals' past experiences and perceptions of pain.

"Not everyone in every culture conforms to a set of expected behaviors or beliefs, so trying to categorize a person into a particular cultural stereotype (e.g., all North Dakota farmers are stoic) will lead to inaccuracies. On the other hand, knowledge of a patient's culture may help you better understand their behavior."[18]

As a native of Kashmir, I have long believed that the Eastern model of reacting to chronic pain is different in many respects from that in the West. For one thing, the culture of machismo is stronger, although gender differences are universal. Also, the influence of religion is more pervasive.

This is especially true in the Hindu faith. In her article "Pain and Suffering as Viewed by the Hindu Religion," Sarah Whitman of the Drexel University Department of Psychology explains: "Suffering, both mental and physical, is thought to be part of the unfolding of karma and is the consequence of past inappropriate action (mental, verbal, or physical) that occurred in either one's current life or in a past life. It is not seen as a punishment but as a natural consequence of the moral laws of the universe in response to past negative behavior. Hindu traditions promote coping with suffering by accepting it as a just consequence and understanding that suffering is not random. If a Hindu were to ask 'why me?' or feel her circumstances were 'not fair,' a response would be that her current situation is the exactly correct situation for her to be in, given her soul's previous action. Experiencing current suffering also satisfies the debt incurred for past negative behavior."[19]

Needless to say, this belief potentially could be problematic for the doctor-patient relationship. If the patient feels that it is his duty to suffer, guiding him along any path of treatment might be difficult. Fortunately, Whitman goes on to say that this is usually not the case: "For Hindus, a first potential challenge may be the feeling of passivity or fatalism that may arise because of karma. A patient can feel hopeless or unable to change things because he feels that things are fixed by karma. Hindu traditions counter this by saying that a person can start in the present moment and go forward, living his life in a positive way by following dharma. If a patient currently experiences pain, change can occur by attending to present appropriate action. If one's present state is a consequence of what has gone before, the urgency of responsible and appropriate action becomes greater."[20]

Buddhists, meanwhile, are trained to take an interior view of pain, looking at it as dispassionately as possible. Darlene Cohen, a San Francisco–based Buddhist and writer, vividly described her struggle with chronic arthritis in her article "Mindfulness and Pain": "Here's where meditation and mindfulness come in. Fully inhabiting my body, despite its devastation, attentive to every little sensation, allowed me to pay close attention to its latent possibilities when they appeared. I lived a half-block from the San Francisco Zen Center when I was just beginning to be able to take walks again, and I used to try to go to dinner there once a week as a treat to myself. Eating a good vegetarian meal with other people.

"Traveling that half-block was my own personal triathlon: walking downhill to the front of the building; climbing the stairs, and knocking on the door with my weak hand. Sometimes I would make it all the way to the steps and not be able to go up them. So I would have to strain all the way back up the hill to my apartment. I asked myself, what is it about my walking that is so tiring? What I called 'walking' was the part of the step when my foot met the sidewalk. From the point of view of the joints, that is the most stressful component of walking. The joints get a rest when the foot is in the air, just before it strikes the pavement. I found that by focusing on the foot that was in the air instead of the foot that was striking the pavement, my stamina increased enormously.

"After making this observation, I never again failed to climb the steps to knock on the front door of the Zen Center."[21]

The prominent religion in my area of Kashmir was Islam, and it was as a Muslim that I was raised. In her exhaustive paper "In Search of a Muslim Pain Principle," Alcira Molina-Ali arrived at the conclusion that the Muslim attitude toward chronic pain is more self-fulfilling than self-destructive. Like Christians and

Jews, Muslims look to their god for healing but are taught to accept divine wisdom if that healing doesn't come.

"Seek treatment for your disease," Muhammad told his disciples. "Allah has not created a disease for which he has not prescribed a cure."[22]

Also like Christianity and Judaism, Islam is a highly diverse religion, ranging from the rather harsh interpretation practiced in Saudi Arabia to the deeply spiritual Sufi strain. Because of this, it presents a range of cultural differences in the medical profession, often unrelated to religion. Muslim women, for example, tend to be extremely modest and uncomfortable undressing in front of male doctors.

In the United States, I practice pain medicine within a primarily Christian context. The problem here, at times, is that a devout Christian's fervent belief that God will heal him or her may lead to depression and a sense of failure if that does not occur. Chronic pain all too often defies a happy ending.

Such insights can be useful in placing various attitudes toward pain in perspective. Nevertheless, many other factors can spawn beliefs and reactions in some individuals that seem to contradict the cultural whole.

As the paper "Cultural Influences on Pain," produced by the International Association for the Study of Pain, points out, "There are differences within cultural and ethnic groups as well as between them. Several factors affect how closely an individual identifies with his or her ethnic or cultural group. These include gender, age, generation, level of acculturation, socioeconomic status (including income, occupation, and education), level of ties to the mother country, primary language spoken at home, degree of isolation of the individual, and residence in neighborhoods made up of one's ethnic group. These factors may mediate the relationship between ethnic background and pain."[23]

Writing in the journal *Practical Pain Management*, Alyson Fincke noted: "Physicians are used to classifying all medical conditions as mild, moderate and severe—or something close to this—in recognition of patient variability. Nevertheless, and amazingly, there is a general attitude that somehow pain is all the same: that there is no variability in severity and . . . pain patients should only be treated one particular way."[24]

Genetics are also emerging as a factor. For example, it has been discovered that the same gene that produces red hair and fair skin may also be connected to greater sensitivity to pain.

"Research has shown that people with red hair perceive pain differently than others. They may be more sensitive to certain types of pain and can require higher doses of some pain-killing medications. However, studies suggest that their general pain tolerance may be higher. People with red hair also respond more effectively to opioid pain medications, requiring lower doses.

"People with red hair have a variant of the melanocortin-1 receptor (MC1R) gene. This gene controls the production of melanin, the pigment that gives skin, hair, and eyes their color. The cells that make melanin produce two forms—eumelanin and pheomelanin. People with red hair produce mostly pheomelanin, which is also linked to freckles and fair skin that tans poorly."[25]

Such discoveries lay open a path to a different approach to pain research that dovetails with gene research on other fronts. If genes can be linked with sensitivity to pain, perhaps they can be engineered to behave differently.

"Pain is a complex human trait sculpted by multiple biologic and psychologic systems, each of which involves the influence of numerous proteins throughout the peripheral and central nervous systems, whose effects can be substantially affected by environmental exposures. Therefore, it is inevitable that multiple genes, each with a small individual effect, interact among themselves and with a variety of environmental factors, to influence pain sensitivity and the expression of chronic pain conditions. Twin studies have demonstrated that genetic influences account for approximately 50% of the variance in chronic pain, and the existing data for experimental pain responses show comparable heritability estimates."[26]

Here again, though, most frontline physicians barely have the time to interact with and initiate treatment for their patients, much less take on an extensive and complicated genetic search. At any rate, in most cases that information would fall into the "interesting, but more than I needed to know" category.

Many hospitals now have geneticists on staff, however, and as the process of genetic detection becomes more streamlined, it might soon become a viable option for pain specialists.

The question is, would delving into a patient's DNA—no doubt at significant cost to that patient or his or her insurance carrier—provide any practical basis for better treatment? Perhaps the main point to be gained from this research is that all pain is not the same.

Previously, physicians and nurses tended to universalize the expected amount of pain according to the procedure—an operation to repair a ruptured spinal disc should hurt this much during the recovery phase, gall bladder surgery that much.

These caregivers also can bring their own culture to their work, tending to admire patients who deal with pain stoically and privately dismissing those who don't as "crybabies."

"I wouldn't act like that if it were me," they might say or think.

Maybe not, but that's not the point. For one thing, it must be considered that patients in a hospital setting have a lot of time to think about how much they are hurting. TV and reading might help for a while, but it's hard for them to forget that their pain has crawled into bed with them.

Effectively alleviating a patient's pain also can be in the self-interest of health-care professionals, since those in pain are often harder to deal with.

"Inadequately managed pain can lead to adverse physical and psychological patient outcomes for individual patients and their families. Continuous, unrelieved pain activates the pituitary-adrenal axis, which can suppress the immune system and result in postsurgical infection and poor wound healing. Sympathetic activation can have negative effects on the cardiovascular, gastrointestinal, and renal systems, predisposing patients to adverse events such as cardiac ischemia and ileus. Of particular importance to nursing care, unrelieved pain reduces patient mobility, resulting in complications such as deep vein thrombosis, pulmonary embolus, and pneumonia. Postsurgical complications related to inadequate pain management negatively affect the patient's welfare and the hospital performance because of extended lengths of stay and readmissions, both of which increase the cost of care."[27]

Additionally, pain is often fluid. We've all had instances where it comes and goes, depending on a variety of factors. Nor does the duration of a medication's effectiveness play out the same for everyone.

"It shouldn't hurt that much," may be the knee-jerk reaction to a patient's complaint, but it's rarely productive.

Insurance companies are another area in which the informing culture tends to lag behind the pace of medical research. These entities may seem impersonal and robotic, but their policies are, after all, determined by human beings who may well bring outdated ideas to the table. One debate currently raging on that front involves whether alternative methods of treating pain can be covered and to what extent.

"Unlike pain practitioners, health care insurers in the United States are not expected to function according to a system of medical ethics. Rather, they are permitted to function under the business 'ethic' of cost-containment and profitability. Despite calls for balancing the disparate agendas of stakeholders in pain management in a pluralistic system, the health insurance industry has continued to fail to take the needs of suffering chronic pain patients into consideration in developing and enacting their policies that ultimately dictate the quality and quantity of pain management services available to enrollees.

"In its efforts to reduce costs, managed care has failed to conceptualize pain as a biopsychosocial phenomenon. Instead, these companies tend to perceive pain as a financial perturbation and inadequately treat it in the cheapest possible manner, i.e., with medications—irrespective of the . . . complications associated with their use."[28]

Thus, a conflict often arises between what the physician feels is the best course of treatment and what the patient can afford.

Chronic pain is a particular sticking point for insurance companies because it often fits into the category of "preexisting conditions." The problem is that those "conditions" can't always be identified to the satisfaction of insurance underwriters, especially when a case of chronic pain seems to come out of nowhere.

The generally conservative bent of those who drive coverage (or the lack of it) also can make them an impediment to new developments in pain treatment. Medical innovations cannot be assessed and refined unless they are used to treat real-life patients, but those patients are often unable to take advantage of them when the cost must be covered out of pocket.

What emerges is a classic catch-22: some of these new medications and techniques actually could cut medical costs, but fear of the unknown may cause insurance providers to reject them. They want any breakthrough to be fully tested on a clinical basis, which is problematic if chronic pain patients—many of whom have lost their jobs because of their conditions and are barely surviving financially—can't pay for them.

"We need to help healthcare providers find more effective ways to treat their patients. Since this is something most doctors do not study adequately during medical school, it's important to have continuing medical education opportunities to learn about the stigma associated with pain treatment and substance abuse disorders."[29]

Moreover, physicians and other caregivers must be careful not to subscribe too closely to cultural stereotypes. Studies have found that some physicians prescribe medication in different doses depending on gender or even ethnicity.

A good example of how prejudice can worm itself into a physician's attitude was the frequent dismissal of complaints by women diagnosed with "chronic fatigue syndrome." Because females were believed to be prone to overstating physical ailments, this affliction was often marginalized—until it became known as fibromyalgia. Again, Weismann, Gordon, and Bidar-Sielaff: "Even more important than understanding the culture of others is understanding how your own upbringing affects your attitude about pain. We are likely to believe that our reaction to pain is 'normal' and that other reactions are 'abnormal.' Thus, a doctor or nurse from a stoic family may not know how to react to a patient who responds to pain by loud verbal complaints (or discount the pain because of the apparent mismatch between the injury and the verbal response). Even subtle cultural and individual differences, particularly in nonverbal, spoken, and written language, between health care providers and patients, impact care."[30]

Other factors, such as race, ethnicity, or religion, can play a role.

"Racial bias in pain management is well documented. There are studies revealing beliefs that Black patients have less sensitive nerve endings than white patients or that Black individuals have thicker skin than any other race and their blood coagulates more slowly. Further, some articles have claimed that Blacks generally experience less pain and thus need lower doses of—or often no—pain medication, and rarely need a narcotic. These ideas may sound absurd, and of course, none of them are valid. Yet almost half of the white medical students and residents in a 2016 University of Virginia study answered 'true' to one or more of these false statements relating to pain assessment. The trainees were also more likely to report lower pain ratings for a Black patient versus a white patient in mock clinical exams."[31]

In other cases, the stereotype may have taken hold that African Americans are more prone to substance abuse. A study by the American Medical Association *Journal of Ethics* on this subject arrived at some interesting conclusions:

- White people are more likely to endanger themselves with the misuse of drugs.
- African Americans and Hispanics were more afraid than were non-Hispanic whites of opioid addiction.
- African Americans and Hispanics were less likely than white people to misuse prescription opioids.
- The overall rate of drug-related deaths was highest among non-Hispanic white people.

Despite this, whites receive more and better pain treatment than African Americans and Hispanics.

- African Americans and Hispanics were less likely than white patients to receive any pain medication and more likely to receive lower doses of pain medication despite higher pain scores.
- They had their pain needs met less frequently in hospice care than did non-Hispanic whites.
- They were more likely to wait longer to receive pain medications in the emergency department than whites.
- Several studies of patients with low back pain found that African Americans reported greater pain and higher levels of disability than whites but were rated by their clinicians as having less severe pain.

- African American and Hispanic veterans with osteoarthritis—particularly African Americans—received fewer days' supply of a nonsteroidal anti-inflammatory drug than white veterans did.
- "Minority" and low-income children were less likely to have oral pain assessed and treated appropriately, especially if they had Medicaid insurance coverage. For example, Hispanic children received 30 percent less opioid analgesia after tonsillectomies or adenoidectomies than white children.[32]

"These findings suggest that clinicians incorrectly believe that Hispanic and African American patients are more likely to abuse drugs than whites and therefore should have less access to them, when in fact they are less likely to do so, and that Hispanic and African American patients experience less severe pain than whites, when in fact they report comparable pain. The findings suggest, in other words, that variations in treatment are based on misconceptions rather than evidence."[33]

Perhaps the blame for this can be partly placed on TV crime shows and movies like *The Wire*, most of which seem to depict the majority of drug dealers as inner-city blacks or Hispanics. This is probably because those who ply their trade on a street corner in a poverty-blighted big-city neighborhood make for a more dramatic storyline than white drug dealers operating via cell phone in some bland suburb.

Complicating the discussion even further are the concepts of "pain threshold" and "pain tolerance." The former is the point at which a person begins to feel pain; the latter indicates how much pain can be felt before it becomes intolerable.

As with most other aspects of pain, these factors are almost impossible to quantify. Women are generally considered to have lower levels of both pain threshold and pain tolerance, but researchers have yet to come up with a definitive explanation.

Moreover, unlike ethnicity, religion, and genetics, pain threshold and pain tolerance can vary exponentially from one day to the next.

A person embarking on an exercise program after a long sedentary period may be painfully sore after the first day, even up to the limits of pain tolerance. With each day, however, the pain lessens until it finally disappears altogether. Does this mean the individual's pain threshold and tolerance have risen, or does that apply only to this particular painful stimulus? No one knows for sure.

Similarly, loud music that might hurt the ears of one person can become less painful with increased exposure. Someone else might be more prone to muscle aches after exertion in cold weather, and most of us tend to feel pain more acutely when fatigued.

There is also a mental component that can move the needle. Many of us have had occasions when the pain of a headache disappears when we become involved in some distraction—until we return our attention to the pain we were feeling. Having worked in hospital emergency rooms, I have seen patients with obviously serious injuries who seem surprisingly free of pain, perhaps due to the adrenalin from a stressful situation.

According to the reports from law enforcement personnel who responded to a car accident involving pro golfer Tiger Woods, Woods at first was trying to stand up and walk even though his left leg was fractured in several places. Yet a year earlier, he was forced to temporarily put his career on hold because of persistent back pain that became intolerable.

One individual has different responses to pain under different circumstances.

Indeed, the degree of pain often does not correlate to the seriousness of the condition. So-called silent heart attacks sometimes occur with minimal discomfort, while another patient's debilitating abdominal pains may turn out to be simple indigestion.

If nothing else, pain is unpredictable. Even the ancient Greeks knew that, although they still struggled to identify a cause.

Interestingly, though, the theory espoused by Aristotle, Plato, and others that pain can be emotional as well as physical has survived long beyond their other suppositions. By this point, the word "pain" has infiltrated the English language as well as the culture—once just a noun, it has mutated into a verb, adverb, and adjective.[34]

We take "pains" to complete a project, often in a "painstaking" manner. We refer to an especially annoying person or an onerous task as "a real pain." We describe a romantic breakup or separation from a job as "painful."

We even sing about pain, with dozens of popular songs that include that word in the title: "King of Pain," "Haven't Got Time for the Pain," "After the Pain," and so on.

Most of us don't have time for the pain, either. But it always has time for us.

Chapter 4

All the Hurtful Things

I feel your pain.

—Bill Clinton, addressing potential voters

The former president didn't mean that literally, of course. But someday, perhaps, advances in technology will allow a physician to access a patient's brain waves in order to feel exactly the same level of discomfort.

But maybe that's not such a good idea. At any rate, unless that occurs, dealing with pain from a medical standpoint will remain very much a guessing game.

For pain is not a finite enemy. It can vary from person to person, day to day, and hour to hour. It can move around the body for no apparent reason. It can stop and start, come and go, turning treatment into a frustrating game of whack-a-mole.

"Exactly how pain works . . . is still largely a mystery. There is no pain center in the brain, no one place where pain signals congregate. A thought must travel through the hippocampus to become a memory, but a pain can surface almost anywhere."[1]

A quick search of the internet using the prompt "first-person pain stories" illustrates that even though most pain is "ordered" by the brain—or, in the case of chronic pain, inflicted by malfunctioning nerves—the varieties of its manifestation are almost endless.

It can be sharp, dull, intermittent, or relentlessly persistent. It can feel like being stabbed by needles, burned alive, or crushed in a vice. And it can ruthlessly assume the central role in any human life.

"You can't find inner peace in that darkness of pain," wrote one cancer sufferer. "I can't emphasize enough that the pain blinds you to all of that, blinds you to all that's positive. I mean the real bad pain . . . just closes you down."[2]

"I couldn't move my body," recalled Angelique, age twenty-four. "I'm a pretty tough gal, so I wouldn't let that stop me, and my mom and I were caravanning the seven hours from my hometown to my university. But the pain was so bad that I began to cry as I waited for a red light to turn green."[3]

Or this, from someone dealing with a kidney stone: "A female nurse told me it's the worst pain a man can ever feel, because a man can't go through labor. I had mine when I was 25; it looked like a coffee grain. Before it passed, I literally thought I was going to die. They gave me a morphine shot right in the vein, and it didn't do a thing. They followed that with Vicodin and I passed out. I woke up a few hours later feeling just OK."[4]

Writes Dutch medical professor Wim Dekkers: "It appears that the meaning of pain is hard to understand from a scientific perspective. Pain is a sensory and emotional experience, the quality of which is difficult to express in words. Pain is a mystery; it cannot be explained as having just a signaling function. It has also an ontological and an existential dimension."[5]

Unfortunately, existential dimensions can't be addressed by basic medical tools. In a 2010 article for the *Journal of Clinical Investigation*, Clifford Woolf provided a cogent explanation of the three main types of pain.

"Pain is actually three quite different things, although we and many of our physicians commonly fail to make the distinction.

First, there is the pain that is an early-warning physiological protective system, essential to detect and minimize contact with damaging or noxious stimuli. Because this pain is concerned with the sensing of noxious stimuli, it is called *nociceptive* pain, a high-threshold pain only activated in the presence of intense stimuli. Its protective role demands immediate attention and action.

"The second kind of pain is also adaptive and protective. By heightening sensory sensitivity after unavoidable tissue damage, this pain assists in the healing of the injured body part by creating a situation that discourages physical contact and movement. This pain is caused by activation of the immune system by tissue injury or infection, and is therefore called *inflammatory* pain; indeed, pain is one of the cardinal features of inflammation. While this pain is adaptive, it still needs to be reduced in patients with ongoing inflammation.

"Finally, there is the pain that is not protective, but maladaptive, resulting from abnormal functioning of the nervous system. This *pathological* pain, which is not a symptom of some disorder but rather a disease state of the nervous system, can occur after damage to the nervous system (neuropathic pain), but also in conditions in which there is no such damage or inflammation (dysfunctional pain). Conditions that evoke dysfunctional pain include fibromyalgia, irritable bowel syndrome, tension type headache, temporomandibular joint disease, interstitial cystitis, and other syndromes in which there exists substantial pain but no noxious stimulus and no, or minimal, peripheral inflammatory pathology."[6]

Both nociceptive and neuropathic pain can be further divided into "acute" and "chronic" categories. In the case of the nerve condition that causes radicular pain, often manifested as "sciatica," a single affliction can qualify in both of those categories.

"Radicular pain occurs because of irritation of a sensitized nerve root. This pain is typically lancinating in quality, shooting down the leg like an electric shock. The patient has difficulty localizing the pain."[7]

Pain in and of itself often tells a physician very little about its cause. Whereas physical trauma is generally obvious—if you have a broken leg, your leg will hurt; a concussion might spawn headaches—chronic pain can lie dormant for months or years, as I discuss in chapter 6. Sometimes, a patient's pain disappears altogether right before a doctor's office visit, the way a malfunctioning automobile stops making that funny noise when you finally take it to a mechanic.

In his book *The Greatest Show on Earth*, Richard Dawkins wrote, "Pain, like everything else about life, we presume, is a Darwinian device, which functions to improve the sufferer's survival. Brains are built with a rule of thumb such as, 'If you experience the sensation of pain, stop whatever you are doing and don't do it again.' It remains a matter for interesting discussion why it has to be so damned painful.

"Theoretically, you'd think, the equivalent of a little red flag could painlessly be raised somewhere in the brain, whenever the animal does something that damages it: picks up a red-hot cinder, perhaps. An imperative admonition, 'Don't do that again!' or a painless change in the wiring diagram of the brain such that, as a matter of fact, the animal doesn't do it again, would seem, on the face of it, enough. Why the searing agony, an agony that can last for days, and from which the memory may never shake itself free? Perhaps grappling with this question is evolutionary theory's own version of theodicy. Why so painful? What's wrong with the little red flag?"[8]

Neuroscientist Patrick Wall echoed Dawkins's complaint, singling out cancer pain as the "apogee of pointlessness. Most cancers don't cause pain in their early stages when it might usefully alert us to take remedial action. Instead, all too often, cancer pain becomes evident only when it is too late to be useful."[9]

Wall's observations came from the heart—he was dying of prostate cancer at the time.[10]

Of course, pain isn't always all-consuming. We might complain about it briefly, then pop an aspirin or two, and feel it slip away. Or our brains can become occupied with other priorities as the pain signals recede into the background. Sometimes, a small pain we take with us to bed is gone in the morning.

Conversely, debilitating pain doesn't have to come from some dramatic event or biological outrage.

Although a toothache is generally not life-threatening, a dysfunctional molar or incisor can produce some of the most intense pain known to man or beast. Dentist Michael Sinkin explains it this way on his website:

"Severe toothache can be a harrowing experience and is in many ways unique from your body's other aches and pains. The intensity of tooth pain can be extraordinary, with severity rivaling true neuralgia—intense neurological pain of almost unparalleled proportions.

"A painful tooth is literally *in your head*. That fact offers you little opportunity to find a comfortable position to neutralize the waves of discomfort. Compared to a painfully sprained foot which you can elevate and use ice packs to get some sort of reprieve, for example, your teeth have an abundance of neural connections to pain centers in your brain. This seems to amplify the noxious distress signals."

Even worse, those distress signals can ambush the tooth's owner. Many of us have had the experience of procrastinating over a minor toothache ("I'm going to have to call my dentist one of these days"), only to have it suddenly attack with latent ferocity.

"For all that makes your teeth especially sensitive to painful stimuli," Sinkin added, "they are also much like any other part of the body. Namely, they can experience transient discomfort that can dissipate almost as quickly as it arises. Aches and pains are a part of an active lifestyle (at least for those of us over 40!), so why should teeth be different? I'm sure you're familiar with the sudden wince when you bite into something unexpectedly hard or the piercing jolt when you chew ice or take too big a mouthful of ice cream. Equally familiar is the agony of a stubbed toe. In most cases, no real damage has been done, but the painful sensation is no less real."

Even so, he said, "Teeth do not respond to trauma in the same way as other parts of your body. When you stub your toe, the injury is mitigated by a robust blood supply that aids in the inflammatory and healing response. Swelling is a natural part of this process, which in time will recede, in large part because of the increased circulation to the area.

"But because a tooth is a solid closed container, the blood supply is restricted and confined. Suppose the injurious insult, such as a broken filling, a chipped tooth, or early to moderate decay, is not too severe. In that case, your dental pulp may be able to respond adequately to the challenge and maintain its vitality.

"On the other hand, there is extreme tooth trauma. Your tooth's circulatory system is not flexible enough to cope with the noxious threat caused by severe physical damage, deep decay, gum disease, etc. Severe trauma can lead to cell death, pulp necrosis, and even abscess formation."

At least a toothache sufferer can pinpoint the source of his or her discomfort. Under the medical "big tent" of arthritis, however, symptoms can vary from "aching, dull, grinding, hot, or throbbing."[11]

As Chicago pain specialist Norman Harden explains: "The pain response . . . just offers us a general impression but not much information about location, intensity, and even seriousness. Very little is known about how pain is processed in the brain. Nevertheless, we know that the brain is very active in processing pain. The pain system is pervasive and redundant in terms of impact on the nervous system but it is not very discrete. In other words, it's qualitative but not quantitative."[12]

The mechanism that informs us of pain is designed to react with incredible speed. By the time the first impulses reach the dorsal horn, seconds have already elapsed, and any delay in relaying that message further could be catastrophic. However, because the nervous system has planted sensors throughout the body, these warning signals can be vague in terms of location.

Think of a smoke alarm, a tool that quickly alerts you to the presence of smoke but leaves it to you to discover its origin. When something has gone wrong inside the body, the answer to the question "Where does it hurt?" often can be misleading.

The pain from spasming stomach muscles, for example, sometimes can be felt in the lower back. In such cases, a patient seeking treatment might say: "I'm pretty sure I've got a kidney stone."

The discomfort from gallstones can settle into the shoulder blade. An angry sciatic nerve can send urgent pain signals along its entire length, from back to leg.

Perhaps the most common example of this "misdirection" occurs when arm pain becomes a heart attack symptom.

"Pain in the arm was mentioned in Heberden's first description of angina pectoris, 'sometimes there is joined (with the chest pain) a pain about the middle of the left arm' (1772). Herrick (1912) also referred to arm pain in his first paper on coronary occlusion, the earliest in the English language. When the arm and chest pain of ischemic heart disease occur at the same time, no diagnostic problem arises. But pain confined to one or both shoulders or arms may be misleading. Its real meaning may be missed even in well-informed circles.

"Not long ago, a professor in an eminent medical school died suddenly from what was shown at autopsy to be a cardiac infarction. For the previous few weeks, he had been having physiotherapy for pain in both shoulders and arms which came on only when he walked. The person with anginal pain mainly in the arms does not associate it with his heart and usually keeps his trouble to himself until he feels the pain in the chest also."[13]

This might seem strange, but the reason for the connection between arm pain and heart attack is even stranger.

It turns out that certain nerves carrying distress signals from the heart to the brain also do the same for the arm. Thus, the brain might be confused about the source of the signal and simply respond to both.

Nor is that the only glitch in the human body's mostly ingenious pain alert network.

Whenever there is a break in the skin, millions of neutrophils—the emergency responders of the immune system—rush to that spot, ready to beat back any possible infection.

"They descend on the scene of the injury like a horde of microscopic barbarians and generally beat the tar out of any microscopic invaders stupid enough to try to crash the party that is you. Neutrophils also destroy a fair bit of healthy tissue in the process. They take *no chances*.

"This is all a normal part of inflammation, and it all makes perfectly good sense. As much as the pain is nasty, this kind of pain makes sense: it's a fair trade. I'm quite willing to put up with the pain, knowing that it's for a good cause.

"But imagine if your local fire department hosed down your house when there *was no fire*. What if they had no concept of a 'false alarm'?

"*Neutrophils have no concept of a false alarm*. None at all.

"Internal injury, sterile tissue damage, any injury where there is exactly zero risk of infection—causes *exactly the same reaction*. Neutrophils rush to the scene and start doing their thing. They attack and kill any cells in the area—ours included—*just in case*. Better safe than sorry, you know! Except it's *not* better, not here, not in this situation."[14]

So what afflictions of the human body hurt the most? Quite naturally, it depends upon who you ask because the answer normally reflects personal experience.

Among the front-runners are kidney stones, gallstones, shingles, reflex sympathetic dystrophy, postherpetic neuralgia, Lyme disease, arthritis, and many types of cancer. Interestingly, women in one survey ranked labor pains further down the list than one might expect, trailing migraine headaches, kidney stones, gallstones, fractures, root canals, and urinary tract infections.

Of course, childbirth pain itself is a variable. Some women report almost painless deliveries, others get through the process only with the aid of robust labor pain medications.

Given the fact that most of us grow up with mothers, the dark accounts of death during childbirth would seem to be overblown. However, not every woman in today's world has the advantage of medical attention during labor. That was the case in Kashmir, where I grew up. And according to a 2016 BBC documentary, childbirth is still not something to be taken for granted.

"Giving birth can be a long and painful process. It can also be deadly. The World Health Organization estimates that about 830 women die every day because of complications from pregnancy and childbirth, and that statistic is actually a 44% reduction on the 1990 level.

"'The figures are just horrifying,' says Jonathan Wells, who studies childhood nutrition at University College London in the UK. 'It's extremely rare for mammalian mothers to pay such a high price for offspring production.'"[15]

In 1960, anthropologist Sherwood Washburn developed the theory that evolution had narrowed the female pelvic bone to accommodate an upright posture while simultaneously increasing the size of the human brain, therefore making infants a tighter fit in the birth canal. This supposition has been debated ever since.[16]

Birth pain is unique in a number of ways, varying not only in intensity but duration. Some children take the express route, others stretch the mother's labor out for long hours, sometimes even days. The pain often comes in waves, especially as the process nears its climax, and the point at which the baby's head emerges is often referred to as "the ring of fire."[17]

"Labor is generally divided into three stages, and the pain associated with each stage has its own source. Beginning from the start of regular uterine contractions until the completion of cervical dilatation, the first stage is associated with visceral pain and is due to the contractions and dilation of the uterine segment and cervix. The second stage leads on from this point up to the completion of delivery, and the pain is primarily somatic as a result of the fetus in the birth canal, causing distention and tearing of the vaginal and perineal tissues. The third stage of delivery is the postpartum stage."[18]

Needless to say, the majority (roughly 60 percent) of women giving birth do so with some chemical assistance. Even those determined to make it "natural" (without medication) often change their minds when the labor pains begin in earnest. The most popular method is the 'epidural,' named for the area of the lower spine into which it is injected. In cases of extreme pain, a "spinal block" can be administered that temporarily shuts down feeling in the lower body.

Most of the drugs currently used in this process are local anesthetics or opioids, which once again raises questions regarding the wisdom of intervening in a natural event.

Although it is generally accepted that these medications do not harm the baby (except for sometimes lowering its blood pressure), one recent study indicated that fentanyl, in particular, could be detected in the bloodstream of some infants after birth—not in significant amounts, but enough cause concern. Opioids are also quick to wear off, which can complicate things for mothers and obstetrician alike.

Meanwhile, there is the matter of oxytocin, a naturally produced hormone that has earned the nickname "the hormone of love." Credited for influencing sexual attraction, it also aids in triggering contractions during labor, produces a calming effect, helps create an immediate bond between mother and newborn, and plays a role in jump-starting breastfeeding.

Scientists have created a synthetic version of this hormone used to alleviate labor pain, which mimics many of these qualities—but not necessarily all of them. In addition, the use of this hormone may reduce the risks of C-section and pregnancy-related high blood pressure.

The list of diverse and different pain types goes on. Cancer pain is a category of its own—and, according to a press release from the Mayo Clinic, is often undertreated.

"Many factors can contribute to that, some of which include:

- Reluctance of doctors to ask about pain or offer treatments. Health-care professionals should ask people with cancer about pain at every visit. Some doctors don't know enough about pain treatment. In that case, request a referral to a palliative care or pain specialist.

  Given current concerns about opioid use and abuse, many doctors might be reluctant to prescribe these medications. Maintaining a close working relationship with your cancer specialists is essential to the proper use of these medications.
- Reluctance of people to mention their pain. Some people don't want to "bother" their doctors, or they fear that the pain means the cancer is worsening. Others worry their doctors will think of them as complainers, or they can't afford pain medications.
- Fear of addiction to opioids. The risk of addiction is low for people with advanced cancer who take pain medications as directed for cancer pain.

  You might develop a tolerance for your pain medication, which means you might need a higher dose to control your pain. Tolerance isn't addiction. If your medication isn't working as well as it once did, talk to your doctor about a higher dose or a different drug. Don't increase the dose on your own.

- Fear of side effects. Some people fear being sleepy, being unable to communicate, acting strangely, or being seen as dependent on medications. You might have these side effects when you start taking strong pain medications, but they often resolve once your doctors find the correct level of pain medications for you and once you achieve a steady level of pain medicine in your body."[19]

Another member of the pain family, the migraine headache, comes with its own puzzle: if the brain has no pain receptors, why do we get headaches?

Physician Allan Basbaum explains, "The brain itself does not feel pain because there are no nociceptors located in brain tissue itself. This feature explains why neurosurgeons can operate on brain tissue without causing a patient discomfort, and, in some cases, can even perform surgery while the patient is awake.

"Headaches, however, are a different story. Though your brain does not have nociceptors, there are nociceptors in layers of tissue known as the dura and pia that serve as a protective shield between the brain and the skull. In some situations, chemicals released from blood vessels near the dura and pia can activate nociceptors, resulting in headaches, such as migraines. Increased blood flow can also trigger a migraine, which is why migraines are considered vascular headaches. Migraine headaches are often throbbing and are accompanied by hypersensitivity to light, sound, and touch."[20]

Indeed, sufferers have described cluster migraines as "feeling as though you are giving birth through your eyeball, but without the prize of a baby when you are done."

In an article for *Science News*, Karen Bannan described one migraine hostage and the drug that helped to free her.

"Hayley Gudgin of Sammamish, Wash., got her first migraine in 1991 when she was a 19-year-old nursing student.

"'I was convinced I was having a brain hemorrhage,' she says. 'There was no way anything could be that painful and not be really serious.'

"She retreated to her bed and woke up feeling better the next day. But it wasn't long until another migraine hit. And another. Taking a pill that combines caffeine with the pain relievers acetaminophen and codeine made life manageable until she got pregnant and had to stop taking her medication. After her son was born, the migraines came back. She started taking the drugs again, but they didn't work and actually made her attacks worse.

"By the time Gudgin gave birth to her second son in 1997, she was having about 15 attacks a month. Her symptoms worsened over time and included severe pain, nausea, sensitivity to light, swollen hands, difficulty speaking, vomiting and diarrhea so intense she often wound up dehydrated in the emergency room.

"This made Hayley Gudgin a prime candidate to test a new drug called erenumab. Her first obstacle was her insurance company (erenumab is not cheap), but she managed to convince them that nothing else was working.

"In August 2018, Gudgin received her first monthly injection of erenumab, sold as Aimovig. By the end of September, she was down to one or two attacks a month. 'And the migraines I do get are usually gone within six hours. I don't have to go to the ER or lie in a dark room all day,' she says. 'It's just been life changing.'

"Gudgin injects the drug into her leg once a month using a device similar to an EpiPen. Erenumab is one of four monoclonal antibodies, manufactured proteins that can bind to substances in the body, that have been approved since 2018 by the U.S. Food and Drug Administration to prevent migraines. The antibodies inhibit the action of a neurotransmitter called calcitonin gene-related peptide, or CGRP, either by changing the peptide's shape or attaching to its receptors in the brain.

"'Roughly half of people who took one of the four drugs in clinical trials saw at least a 50 percent reduction in monthly migraines,' says neurologist David Dodick of the Mayo Clinic in Phoenix, who reported the findings at a Migraine Trust International Symposium in October. About a third of patients had a 75 percent drop in migraine days.

"The CGRP-blockers appear to be an improvement over existing preventive treatments, which were developed for other disorders. The newer drugs were designed specifically to target one of the mechanisms that researchers think leads to painful episodes.

"Doctors are embracing the new drugs because they can work better and generally have much fewer side effects than other options. 'It's really beneficial for improving quality of life in our patients with migraines. [The new drugs] don't cause weight gain, sleepiness, brain fog,' says neurologist Nina Riggins, a headache specialist at the University of California, San Francisco."[21]

The rush to produce viable COVID-19 vaccines provided a glimpse into the brave new world of drug research and development. It is a highly competitive world, particularly given the rivalries among pharmaceutical companies, but it is also often symbiotic.

In the twenty-first century, most of the research on pain and pain relief is being done either by university-based science and medical departments, medical centers such as the Mayo Clinic, or "Big Pharma" (plus, in some cases, smaller pharma). All of these come with built-in advantages and disadvantages.

Medical centers and universities are much more likely to seek out a cure, an antidote, or a potential cause for its own sake, simply to expand the communal storehouse of knowledge. Because they are generally funded by government grants and alumni gifts, they are relieved of the necessity of focusing only on discoveries that lead directly to future revenues. This allows them a great deal of flexibility in choosing research topics.

The flip side is that they lack the deep pockets of the major drug companies.

Meanwhile, there are literally thousands of diseases and other health conditions that afflict only a small portion of the general populace. Because the potential for profit is so small in these cases, the pharmaceutical companies generally ignore them. It is, after all, difficult to justify a sizable outlay of research and development money to company stockholders if only a few hundred people will benefit—and, by extension, are likely to pay for the resulting product.

Still, these two exploratory entities do have a vested interest in working together—or at least in keeping apprised of what the other is doing.

The new migraine headache drugs are a good example. Like many other chronic pain conditions, migraines were once considered one of nature's inexplicable curses, an affliction that could never be cured—or even understood—but only survived. All the Hayley Gudgins of the world could hope for was to weather each attack and hope the next one wouldn't be as bad.

That was before a broad-based effort was instituted on many levels to find out more about the brain. As each layer of that incredibly complex organ was peeled back and examined, scientists both private and public realized that virtually everything that happens within our bodies has a rationale behind it, and that rationale can almost always be traced to the brain.

This is when the pharmaceutical companies stepped in, especially when initial research into some common human conditions promised a significant market for a possible new drug. Once university or medical center researchers had opened the door by identifying a cause and effect, only the drug companies had the means to directly connect that discovery with those it could help.

To build that bridge, however, they also need physicians who deal with patients face-to-face and often have the option of deciding which of several similar medications to prescribe. We don't take that responsibility lightly.

As a specialist, I have dealt with pain in virtually all its forms. Three of its manifestations that have particularly impressed me are those from reflex sympathetic dystrophy (RSD), shingles, and postherpetic neuralgia (PHN), none of which generally comes up in casual conversations about things that hurt.

RSD and PHN are undeniably painful. PHN, though not life-threatening in the usual sense, has driven patients to suicide because they couldn't imagine living any longer with pain of that intensity. RSD combines pain with muscle and tissue damage, which can be irreversible. In extreme cases, physicians have decided that the only way to stop RSD pain in a limb is amputation.

The support group America RSD Hope describes the symptoms as "burning pain, as if a red hot poker were inserted into the affected area; also throbbing, aching, stabbing, sharp, tingling, and/or crushing in the affected area (this is not always the site of the trauma). The affected area is usually hot or cold to the touch. The pain will be more severe than expected for the type of injury sustained.

"Allodynia is usually present as well (extreme sensitivity to touch). Something as simple as a slight touch, clothing, sheets, even a breeze across the skin on the affected area can cause an extreme amount of pain to the patient. Pain can also be increased by sounds and vibrations, especially sharp sudden sounds and deep vibrations."[22]

As for PHN, it sometimes appears first in the form of shingles—the lesser of two evils, perhaps, but still a potential nightmare.

In his article "The Biological Aspects of Pain," Dr. Richard W. Hanson wrote, "A large part of this apparent puzzle [of chronic pain] . . . results from a failure to understand the active role played by the brain in pain perception. According to this view, the brain is simply a passive receiver that picks up pain signals that are generated and transmitted directly from the site of the injury. In other words, the brain is seen to work something like a telephone or radio receiver that accurately reproduces whatever messages are sent to it.

"In reality, the brain is not simply a passive receiver, nor is the spinal cord a passive conveyor of pain messages originating in some injured part of the body. Rather, both of these central nervous system structures play an active role in modifying the pain messages that are ultimately registered in the brain. Furthermore, the brain serves as both a receiver and an active transmitter. It can transmit signals that block the experience of pain. It can also significantly magnify the experience of pain out of proportion to the original injury, or it can even generate signals which lead a person to experience pain in a part of the body that is not actually injured."[23]

Like a rock tossed into a pond, pain often spreads out in ripples throughout the body.

"A single painful location on the body soon begets some others. Much of this is the 'overload and overuse syndrome.' To make up for a weak, painful area, joints, nerves, and muscles elsewhere in the anatomy will attempt to compensate and work overtime. Unfortunately, chronic overuse and overload may lead to tissue degeneration at secondary pain sites causing arthropathies, myopathies and neuropathies. Pain patients sometimes develop a more painful secondary site than the primary site. The astute practitioner can often observe physical evidence of overuse, overload, and self-splinting such as the presence of hypertrophy of paraspinal or neck muscles, unilateral furrows of the brow in the persistent headache patient, and calluses on the foot in the patient with back, hip, or leg pains."[24]

The inertia and inactivity that pain sometimes demands takes a toll as well, leading to such side effects as obesity and muscle atrophy. It also affects certain chemicals and hormones in the body.

"Based on emerging research data, it appears that uncontrolled persistent pain may affect about every endocrine system in the body. It has long been observed that acute pain is often accompanied by hypertension and tachycardia, and it is now clear that persistent pain may actually trigger indolent hypertension and tachycardia. Cardiovascular death is a common occurrence among persistent pain patients, likely due to a multitude of factors.

"Immune suppression is present in the persistent pain patient. It is manifested clinically by poor resistance against infections and slow healing of wounds or injuries. Hormonal abnormalities are most likely responsible. At this time, little is known about pain's effects on such hormones as glucagon, thyroid, insulin, growth hormone, estrogen, progesterone, and endorphins, but derangement of any of these—in addition to cortisol, pregnenolone, and testosterone—may adversely affect the immune system. Serum testing of persistent pain patients typically shows a variety of serum immune abnormalities. Opportunistic infection is another hallmark of a suppressed immune system. Persistent pain patients, particularly those with an autoimmune disease such as fibromyalgia or systemic lupus erythematosus, may develop infections such as chlamydia, cytomegalovirus, and herpes. It may be that the same nervous tissue that produces pain is so injured that it becomes fertile ground for viruses that like to invade nerves."[25]

And it gets worse.

"'Not only can chronic pain compromise one's emotional and mental mindset,' says pain specialist Dr. Joanne Witkowski, 'it can also trigger physiologic changes. Some studies suggest that people with chronic pain actually have a smaller hippocampus, the section of the brain responsible for memory formation, organization, and sorting.' Additional research also indicates that chronic pain can create changes in your spinal cord."[26]

Even acute pain can be life-threatening in some cases.

"Extreme pain causes neurogenic shock by overexciting the parasympathetic nervous system. This results in a significant decrease in heart rate [bradycardia], which in turn decreases the pulse and leads to a dangerous drop in blood pressure [shock]."[27]

Aristotle was, indeed, wise when he said that avoiding pain should be a priority, and Richard Dawkins had a legitimate complaint about its intensity. Nevertheless, whether you consider it an ally or a foe, pain will continue to play a major role in human existence.

That's why I'm here.

Chapter 5
Chasing Relief

US presidents have the Secret Service to protect them; heads of state in other countries have military bodyguards. The human body has nociceptors.

Just as the sudden appearance of a man with a gun outside the White House summons presidential defenders at a run, these uniquely specialized neurons react almost instantly to signs of danger. Yet even that analogy falls short because every centimeter of the human body plays host to at least two hundred nociceptors—the rough equivalent of a Secret Service agent posted on every square foot of Washington, DC.

To employ the usual example, if you accidentally put your hand on a hot stove burner, specific nociceptors immediately shout to the brain: "Get your hand off that stove! Now!"

In fact, you probably reflexively jerked your hand away even before you had an instant to think about it.

"Just like all nerve cells," explains scientist and author George Zaidan in a popular TED talk, "they [the nociceptors] conduct electrical signals, sending information from wherever they're located back to your brain. But unlike other nerve cells, nociceptors only fire if something happens that could cause or is causing damage."[1]

And these literal bodyguards definitely have us covered, taking up their vigil wherever there is skin. Or internal organs. Or even our teeth. They maintain their vigilance, in fact, everywhere except inside the tissue of the brain itself.

Researchers have identified three types of nociceptors—mechanical (on guard in case of physical injury), temperature related (the last line of defense against such bodily hazards as frostbite and sunburn), and chemical. They provide an essential service, to be sure, but the downside is pain. Without the unpleasant reminders these sentries convey, you might have left your hand on that theoretical stove until you began to smell burning flesh.

Pain relievers, then, represent a trade-off—in exchange for the elimination or diminution of an unpleasant sensation, they assume the authority required to short-circuit a useful bodily function. In effect, they are telling the nociceptors that the brain doesn't really need to hear about this issue just yet.

Therefore, as a physician, treating pain often means walking a narrow path between necessary relief and a return to normality. If all goes well, the type and dosage of pain drugs can and should be scaled back gradually until none are needed.

Of course, pain sometimes becomes so acute that it can nudge a patient over the edge into shock. At such times, intense drugs might be called for. Generally, prompt attention to the condition causing the pain diminishes the discomfort rather quickly—and with it, the need for extreme drug therapy. The difference, as we see in the next chapter, comes in cases of chronic pain.

We acquire our knowledge about the world around us (and within us) through a messy process that plays out raggedly over time. As part of that process, we struggle to splice together isolated and seemingly disconnected facts and occasionally are forced to backtrack from what turn out to be dead ends. Often, we learn in reverse, first determining that something exists or has a particular effect, then deferring questions like "how?" or "why?" for later researchers to figure out.

For example, the vital connection between the brain and pain has been attributed, over time, to everything from outside demonic forces to the influence of the heart. Gradually, one research breakthrough (or lucky guess) at a time, the explanation moved closer to where it needed to be. In the mid-1600s, Rene Descartes broke new ground, and the discussion shifted again.

Writing in his 1664 *Treatise of Man*, Descartes theorized that the body was similar to a machine, and that pain was a disturbance that passed along nerve fibers until the disturbance reached the brain. This theory transformed the perception of pain from a spiritual, mystical experience to a physical, mechanical sensation, meaning that a cure for such pain could be found by researching and locating pain fibers within the body rather than searching for an appeasement from god. This also moved the center of pain sensation and perception from the heart to the brain.[2]

"Descartes proposed his theory by presenting an image of a man's hand being struck by a hammer. In between the hand and the brain, Descartes described a hollow tube with a cord beginning at the hand and ending at a bell located in the brain. The blow of the hammer would induce pain in the hand, which would pull the cord in the hand and cause the bell located in the brain to ring, indicating that the brain had received the painful message. Researchers began to pursue physical treatments such as cutting specific pain fibers to prevent the painful signal from cascading to the brain."[3]

At the time, this was a brave and unique interpretation. If nothing else, Descartes had moved the conversation to where it belonged—inside the complex human nervous system. His mistake, however, was presenting pain as a unified reaction, always identified and sent to the brain in the same way by the same nerve center. It was almost another 300 years before British researcher Charles Scott Sherrington discovered the role played by the nociceptors.[4]

Once the presence of nociceptors was further confirmed through additional experiments by Sherrington and others, the search turned to discover more information about the process used in triggering the pain reflex.

A later academic paper explains, "The peripheral terminal of the mature nociceptor is where the noxious stimuli are detected and transduced into electrical energy. When the electrical energy reaches a threshold value, an action potential is induced and driven towards the CNS [central nervous system]. This leads to the train of events that allows for the conscious awareness of pain."[5]

In addition to the specialized nature of some nociceptors, these neuron fibers are roughly divided into categories labeled A Delta and C. This delineation has to do with the speed of their response, the nature of the pain, and the thickness of the protective myeline sheath around them (the thicker the sheath, the quicker the response).[6]

A Delta nociceptors bring forth the sharp, immediate pain that causes us to react almost instantly. This pain is usually of short duration, followed by a duller sensation (dispatched by the C receptors) that last longer.[7]

In other words, the A Delta fibers tell us, "This is going to hurt," whereas their C counterparts chime in with, "And it's going to hurt for a while."

As explained on the website *The Brain from Top to Bottom*:

"This time lag is directly attributable to the difference in the conduction speeds of A delta and C fibres: their messages do not reach the brain at exactly the same time. 'Fast pain,' which goes away fairly quickly, comes from the stimulation and transmission of nerve impulses over A delta fibres, while 'slow pain,' which persists longer, comes from stimulation and transmission over non-myelinated C fibres. In relative terms, A delta fibres carry messages at the speed of a messenger on a bicycle, while C fibres carry them at the speed of a messenger on foot. C fibres are estimated to account for about 70% of all nociceptive fibres.

"Myeline, a fatty tissue, serves the same function as tape wrapped around electrical wires. Researchers seeking the cause of multiple sclerosis have discovered that when that outside sheaf is damaged, the fibers within (known as axons) are prevented from using the nociceptors to send messages along to the brain. In the case of MS, those messages are more likely to involve commands to the muscles than pain, but the discovery of this phenomenon raised hope that brain-nervous system communication might be interrupted in other ways."[8]

Entire books have been written and research careers focused on the swift but incredibly complicated journey from nerve ending to the brain. There are variables within nociceptors to account for. A wide variety of chemicals and hormones are automatically dispensed to aid in the pain recognition process. The impulses first travel together to the dorsal horn of the nervous system, which serves the same function as a connecting airport, then branch off to interact with the brain at different places. And all of this takes place in the blink of an eye.

Well, almost in the blink of an eye.

"Light travels at 300 million meters per second, while nerve signals move at a decidedly more stately 120 meters a second—about 2.5 million times slower. Still, 120 meters a second is nearly 270 miles an hour, quite fast enough over the space of a human frame to be effectively instantaneous in most instances. Even so, as an aid to responding quickly, we have reflexes, which means that the central nervous system can intercept a signal and act on it before passing it on to the brain. That's why if you touch something very undesirable, your hand recoils before your brain knows what's going on.

"The spinal cord," in short, "is not just a length of impassive cabling carrying messages between the body and the brain but an active and literally decisive part of your sensory apparatus."[9]

For my fellow medical professionals, there are plenty of sources that explain all this in detail, albeit couched in language too technical for the average reader. Since the goal of this book is to reach and inform a more universal audience, that means keeping it as simple as possible, focusing on two primary questions: why do we hurt, and what can we do about it? A guided tour of the nervous system might be of interest to some people but a bit much for most.

Certainly, though, the intricate dance performed by the brain and nervous system is worthy of awe, if not total understanding.

The brain is always occupied, even during sleep. Yet whatever a person might be thinking, the news of impending pain will almost always claim his or her immediate and rapt attention, like a bulletin scrolling across the bottom of a television or computer screen.

As a universal human problem, pain has never been something to set aside and worry about later. Just imagine the days prior to anesthesia, when intrusive medical operations were carried out on patients who were conscious, grimacing, and sometimes screaming. Obviously, this was also a problem for surgeons confronted with moving targets for their scalpels. (Incidentally, recent research has suggested that the reason most of us yell or scream when inflicted with pain is not to call attention to ourselves, but to instinctively interfere with pain signals traveling to the brain.)[10]

Although some of the substances at the heart of our current pain medications have been known and used for thousands of years, they were initially valued for their euphoric qualities and used primarily in religious ceremonies.

Today, the pleasurable feelings that come with many pain medications are considered secondary to the reduction of pain; in medieval times and before, those priorities were reversed.

References to opium have been found as far back as 3400 BC, the date of a Sumerian clay tablet that clearly depicted the controlled growing of poppies. Still, it is unclear how and when these colorful flowers became more than just eye candy.

Since opium is derived from the dried sap of the plant's seed pod, that means someone in the ancient world decided to either taste the sap oozing from that pod or boil it into a drink. Why that initial impulse to experiment? No one knows, but the opium poppy soon earned the Sumerian name *Huy Gil*, or "the joy plant."[11]

Before long, Sumerian traders introduced it to their Assyrian counterparts, who then passed it along to Egypt (where it was cultivated in massive fields). The joy plant eventually infiltrated the Middle East and medieval Europe and followed the Silk Road to China, where it became a significant part of everyday life. English visitors to China took it home (sparking two opium wars when China tried to cut off the supply), and Chinese immigrants finally brought it to the United States.

Opium had always been used by those in pain, but more as a distraction than an actual intervention. That began to change in the fourteenth century, when the shadow of the black plague moved across Europe, killing one of every three people in its path according to some accounts.

Practical medicine was in its infancy then, and many of the initial attempts at defeating the plague were inspired primarily by superstition and folk legend. Many would seem laughable were they not expressions of desperation.

Black cats were blamed for the disease at one point, triggering a massacre of those animals. Unfortunately, that gave more leeway to the rats that were actually spreading the deadly germs.

In other places, it was recommended that a live chicken be attached to the patient, based on the theory that the chicken would "draw out" the pestilence.

From amid this chaos came a small army of "plague doctors," perhaps history's first medical specialists, most of whom brewed up their own versions of a "theriac," the Greek word for a universal antidote.

"Different kinds of theriacs were . . . produced in antiquity but the most celebrated was perhaps that invented by Andromachus, physician to the Roman Emperor Nero, in the first century AD. Andromachus came from the island of Crete where 'botanical men' in the service of the Emperor collected herbs and placed them in knitted vases, which were sent not only to Rome but also to other nations. Andromachus' vast knowledge of botany helped him 'to provide mankind with the necessary medicines.' He claimed that his formula for his Galeni Theriaca (tranquility theriac) was an improvement . . . because it contained some 64 ingredients and was enriched with the flesh of a viper and a much greater quantity of opium."[12]

Vipers were hard to find, but opium was plentiful.

The same formula, or some semblance of it, was later produced by the Greek physician Galen, who popularized it. The concept then "survived into medieval Europe in the trade that developed in theriacs, most notably in Italy, where theriacs became known as the Venice Treacle, an official preparation that carried the republican seal. Its legacy is even apparent in French and German pharmacopeias of the 19th century. Whether as a universal panacea or just an addictive preparation thanks to opium, the theriac's influence extended far beyond its origins in antiquity."[13]

Among other things, Galen claimed his theriac could protect against poisons and snakebite.

"In the royal court, Galen prepared his theriac and wrote about various theriac compounds in his books *De Antidotis I*, *De Antidotis II*, and *De Theriaca ad Pisonem*. The basic formula consisted of viper's flesh, opium, honey, wine, cinnamon, and then more than 70 other ingredients. The final product was supposed to mature for years and was administrated orally as a potion or topically in plasters. Galen claimed that his theriac drew out poisons like a cupping glass and could divide the tissue of an abscess more quickly than a scalpel. The preparation was taken daily by the Emperor Marcus Aurelius to protect against poisons and to aid in ensuring good general health.

"But Galen did not just administer his theriac, he also experimented with it on animals. In *De Theriaca ad Pisonem* he describes how he took roosters and divided them into two groups: in one group he gave the theriac and in the other group he did not. Then he brought both groups into contact with venomous snakes; Galen observed that the roosters who had not been given the theriac died immediately after being bitten, whereas those who had been given the theriac survived. Moreover, he points out that this experiment could be used in cases where someone wants to make sure whether a theriac is in its natural form or has been adulterated. Alongside this work, Galen wrote about the effect of his theriac on individual patients. In one passage in *De Theriaca ad Pisonem*, he gives an illuminating account of using his theriac to treat jaundice caused by snakebite."[14]

Eventually, the black plague ran its course, and opium then returned to its roots as a mind-altering vessel.

"Recreational use of the drug was taken up enthusiastically by the citizens of the Persian Empire during the late medieval period. Rulers of the Mughal Empire formed opium habits by eating it; Emperor Jahangir was so inebriated on the drug and wine it left him incapable of ruling. His wife had to fill his role. In Turkey, its use was so widespread, it was said: 'There is no Turk who would not buy opium with his last penny.'"[15]

In 1676, physician Thomas Steadman universalized Jahangir's method of addiction by publishing his recipe for laudanum. Whereas earlier opium users had merely used it in conjunction with wine, laudanum combined the two in one powerful liquid.[16]

"Laudanum was a 10 percent solution of opium powder in alcohol, widely used to treat everything from pain and insomnia to female disorders. It was even used to quiet crying babies. The name was coined from the Latin *laudare*, meaning 'to praise' by the 16th century Swiss-German physician Paracelsus, best known for his dictum 'only the dose makes the poison.' It was Paracelsus who discovered that the active ingredients in opium were more soluble in alcohol than water, however, his laudanum differed from the version used in the Victorian era. In addition to opium it also contained powdered gold and pearls."[17]

This new version of opiate relief did indeed contain "gold and pearls" for the companies that produced it. Already well established in England, the lure of opium in all its forms was carried to America by the first wave of immigrants.

"Opium's history in the United States is as old as the nation itself. During the American Revolution, the Continental and British armies used opium to treat sick and wounded soldiers. Benjamin Franklin took opium late in life to cope with severe pain from a bladder stone. A doctor gave laudanum, a tincture of opium mixed with alcohol, to Alexander Hamilton after his duel with Aaron Burr."[18]

The American Civil War, tragic and brutal as it was, required a flood of opium to deal with combat wounds, rampantly contagious diseases, and the postwar agonies of soldiers who brought the lingering pain from amputations home with them. Many of them became opium addicts.

The first popular derivative of opium was discovered by an obscure, largely uneducated German physician's assistant named Friedrich Wilhelm Serturner after the turn of the nineteenth century. By immersing crude opium in ammoniated hot water, Serturner produced a yellowish-white crystalline compound.

"He first tested this compound on a few dogs, which resulted in their death. He then tested smaller doses on himself and some boys and found that the effects were pain relief and euphoria. He also noted that high doses of the drug could lead to psychiatric effects, nausea, vomiting, depression of the cough, constipation and slowed breathing. Pain relief with the use of this compound, however, was ten times that experienced with opium use.

Serturner named his compound morphine, after the Greek god of dreams, Morpheus."[19]

Meanwhile, the search was on to find a safe way to render patients unconscious (or, at least, relatively unfeeling) during surgery. William T. G. Morton, a Boston dentist, had been experimenting with nitrous oxide when he happened to hear an 1844 lecture by Harvard University chemistry professor Charles Jackson, who had discovered that sulfuric ether could "render a person unconscious or even insensate."[20]

After a period of research and testing, Morton used that organic compound in 1846 to anesthetize a patient who was having a tumor removed from his neck.[21] Morton named his discovery "Letheon" for the river in Greek mythology said to erase painful memories. It wasn't until 1902 that a Chicago physician, Mathias J. Seifert, introduced the term "anesthesia."

Even at that point, however, it appeared that true pain relief had become a matter of all or nothing. Inducing unconsciousness during an operation was obviously welcomed under those circumstances, but just as obviously impractical for nonsurgical use.

So researchers kept grasping for the prize, and new drugs continued to emerge from the netherworld of beakers and solvents and laboratory mice. Codeine appeared in 1830. Heroin was synthesized in 1874 under the illusion that it would be less addictive than opium, and when that proved a miscalculation, methadone came along in 1937 with the promise of being less addictive than heroin.

Eventually, a parade of new drugs began to emerge not from medical researchers, but pharmaceutical companies. Vicodin arrived in 1984, OxyContin in 1995, Percocet in 1990.

This left another niche to be filled: that of combating relatively low-level pain (headaches, muscle aches, etc.) for which opioid drugs would be overkill. The Bayer Company took the first giant step along that parallel track in 1899, announcing the development of aspirin. That product ruled the "small pain" market for more than a half-century before the emergence of Tylenol (1955) and acetaminophen (1956). Ibuprofen made its debut in 1962.

"The ancient Egyptians used willow bark as a remedy for aches and pains," said Diarmuid Jeffreys, author of *Aspirin: The Remarkable Story of a Wonder Drug*.[22] They didn't know that what was reducing body temperature and inflammation was the salicylic acid.

"Hippocrates, the Greek physician who lived from about 460 to 377 BC, wrote that willow leaves and bark relieved pain and fevers. It wasn't until thousands of years later that people began to isolate the key ingredients of aspirin.

"An 18th-century clergyman, Edward Stone, rediscovered aspirin, in effect, when he wrote a report about how a preparation of powdered willow bark seemed to benefit 50 patients with ague and other maladies.

"In the 1800s, researchers across Europe explored salicylic acid. French pharmacist Henri Leroux isolated it in 1829, Roueché writes. Hermann Kolbe discovered synthetic salicylic acid in 1874, but when administered often in large doses, patients experienced nausea and vomiting, and some even went into a coma. A buffer was needed to ease the effects of this acid on the stomach.

"The aspirin we know came into being in the late 1890s in the form of acetylsalicylic acid when chemist Felix Hoffmann at Bayer in Germany used it to alleviate his father's rheumatism, a timeline from Bayer says. Beginning in 1899, Bayer distributed a powder with this ingredient to physicians to give to patients. The drug became a hit and, in 1915, it was sold as over-the-counter tablets."[23]

British pharmacologist John Vane broke new ground on the research side in the early 1970s with several discoveries that eventually earned him a Nobel Prize in 1982. Most significantly, he examined how aspirin neutralized the effects of prostaglandins.[24]

Prostaglandins are hormone-like substances that govern several important processes in the body. They also come into play when the body is under attack. In 1971 Vane showed that acetylsalicylic acid, a substance found in pain-relieving and fever-reducing medications like aspirin, works by inhibiting the formation of prostaglandins.[25]

Adds George Zaidan in his TED talk, "How Do Pain Relievers Work?": "But the pain threshold isn't set in stone. Certain chemicals can tune nociceptors, lowering their threshold for pain. When cells are damaged, they and other nearby cells start producing these tuning chemicals like crazy, lowering the nociceptors' threshold to the point where just touch can cause pain. And this is where over-the-counter painkillers come in. Aspirin and ibuprofen block the production of one class of these tuning chemicals, called prostaglandins.

"Let's take a look at how they do that. When cells are damaged, they release a chemical called arachidonic acid. And two enzymes called COX-1 and COX-2 convert this arachidonic acid into prostaglandin H2, which is then converted into a bunch of other chemicals that do a bunch of things, including raise your body temperature, cause inflammation, and lower the pain threshold. Now, all enzymes have an active site. That's the place in the enzyme where the reaction happens.

"The active sites of COX-1 and COX-2 fit arachidonic acid very cozily. As you can see, there is no room to spare. Now, it's in this active site that aspirin and ibuprofen do their work. So, they work differently. Aspirin acts like a spine from a porcupine. It enters the active site and then breaks off, leaving half of itself in there, totally blocking that channel and making it impossible for the arachidonic acid to fit. This permanently deactivates COX-1 and COX-2. Ibuprofen, on the other hand, enters the active site, but doesn't break apart or change the enzyme. COX-1 and COX-2 are free to spit it out again, but for the time that that ibuprofen is in there, the enzyme can't bind arachidonic acid, and can't do its normal chemistry. But how do aspirin and ibuprofen know where the pain is? Well, they don't. Once the drugs are in your bloodstream, they are carried throughout your body, and they go to painful areas just the same as normal ones.

"So that's how aspirin and ibuprofen work."[26]

As for the effects of opioid drugs, the website *ATrain: Continuing Education for Health Care Professionals* provides a relatively straightforward explanation:

"There are several neurotransmitters involved in pain signals, the main ones being *glutamine* and *substance P.* When noxious stimuli trigger the primary neuron through the skin or muscle, the message is relayed by a secondary neuron to the spinal cord's dorsal root ganglion and toward the brain for interpretation. These chemical neurotransmitters are relayed to the thalamus in the brain and then onto the limbic system for an emotional response. Ideally, the message to the limbic area of the brain promotes learning so as to avoid the cause of the noxious substance in the future.

"Opioids inhibit pain signals at multiple areas in this pathway. They affect the brain, the spinal cord, and even the peripheral nervous system. Opioids work on both directions of messages in the nervous system, including the ascending pathways in the spinal cord, which they inhibit, and the descending pathways, by which they block inflammatory responses to noxious stimuli.

"Our bodies have three receptors called mu, kappa, and delta, that can be activated by opioid agonists like morphine, hydrocodone, or heroin. When mu receptors are activated, dopamine, the natural brain chemical for pleasure, is also increased. Pleasurable feelings are experienced as inherently worth repeating, which drives the user to repeat the drug use.[27]

"Opioid receptors are found on both the primary and secondary neurons, and when an opioid binds to these receptors no other pain signals are sent up to the brain—making opioids very effective against pain. In the brain, opioids cause sedation and decrease the emotional response to pain. Heroin, like morphine, passes through the liver and then is released back into the blood, where it crosses the blood-brain barriers. Heroin is converted to morphine, where it connects with mu receptors so fast that heroin is three times more potent than morphine."[28]

The purpose of these drugs, however, is not to hijack the body's natural pain-fighting abilities forever, just to deliver "more bang for the buck" in the short term. Continual use over time not only lessens their effectiveness (encouraging the patient to take higher and more frequent dosages) but also negatively affects respiration, heart rate, and even the brain's white matter.

This brings us to the dilemma of chronic pain.

Chapter 6

How Chronic Pain Changed the Game

Chronic pain is where the medical profession went off the rails.

Over many generations, one solved mystery at a time, a plausible scenario finally emerged as to how and when pain affected the human body. The process, it turned out, was largely driven by strategically placed nociceptors that sensed impending tissue damage and passed the information along to the brain, which then translated this news into actual pain.

The first painkillers were developed before we knew that. Once the presence and function of the nociceptors were discovered, however, we realized that certain medications temporarily muted these tiny sentinels so that their message was never delivered. It was reminiscent of that old philosophical dilemma about the tree falling in the forest—if the brain isn't told about our pain, does the pain actually exist?

Although the explanation of this process was complicated and laced with arcane medical terms, it made sense at its core, and it obviously appeared to work most of the time.

But what about when it didn't? What if deactivating the nociceptors didn't keep the pain from happening? Or what if X-rays, CT scans, and other tests showed that an injury had healed completely, but the brain kept sending out messages to the contrary?

That's chronic pain.

With acute pain, communications from the nerves generally fly straight and true to the brain. Their purpose is to alert us to injury and illness. They are on our side.

Chronic pain, on the other hand, can act like a malfunctioning GPS device. Its messages can sometimes wander awry, sending the alarm to somewhere in Kansas instead of Chicago. Or the message gets caught in a ceaseless viral loop, declaring the presence of pain over and over again.

In her vivid and perceptive memoir and examination of chronic pain, *The Pain Chronicles*, Melanie Thernstrom provides this analogy: "Imagine a home security alarm system that is first triggered by a cat, then a breeze, and then, for no reason, begins to ring randomly or continuously. As it continues to ring, it triggers other noises in the house: the radio and television start to blare; the oven timer dings; the doorbell buzzes repeatedly, and the phone rings maniacally even though no one has placed a call. This is neuropathic pain."[1]

Moreover, the stridency of the alarm is often significantly out of proportion to the damage that triggered it. Some patients with reflex sympathetic dystrophy or fibromyalgia find their torment unleashed not from a broken bone or severe trauma, but simply by reaching down to pick up a book or twisting slightly while exiting a car. For them, the punishment far exceeds the crime.

"If significant tissue damage has occurred, or if there has been a prolonged or particularly intense activation of a primary nociceptor, it will become sensitized. Sensitized nociceptors can be activated by modern stimuli that normally don't produce pain. One common example of sensitized nociceptors is the agony produced by bath or shower water on sensitized skin."[2]

So how does a physician handle a complaint like "It hurts when I take a shower"?

In essence, we are taught medicine on a cause-and-effect basis, which starts with first comparing a patient's symptoms to a database of possible causes. When we find a place where symptoms and conditions seem to match, we recommend a course of treatment that usually includes medication. If no improvement results, we then move back to the data and search for a plan B.

Along the way, a steady drumbeat of technological advances has removed much of the guesswork from this process. Modern blood tests can not only reveal the percentage of sugar and other substances, but also alert to various types of cancer. X-rays and CT scans act like microscopes trained on the body to reveal its innermost secrets, which is revealed in black and white and numbers, leading the physician down a clearly delineated path to treatment.

Except when it doesn't. Dealing with chronic pain is, in many ways, a throwback to the old intuitive style of doctoring, forcing us to admit: "Something seems to be broken here, but I'm not sure what it is."

Adding to the problem is that such conditions sometimes don't show up until months or years after treatment for the original injury or illness is finished.

For instance, the medical profession's remarkable success in treating and curing cancer over the past twenty years is certainly good news. The downside is that it has created yet another opportunity for chronic pain gremlins.

"A report by researchers from Mount Sinai, Memorial Sloan-Kettering Cancer Center, University of Virginia, and the American Cancer Society finds that about one-third of cancer survivors (34.6%) reported having chronic pain. That is almost double the rate in the general US population. The report was published on June 20, 2019, as a research letter in *JAMA Oncology*.

"Chronic pain is one of the most common long-term side effects of cancer treatment and has been linked with lower quality of life, less likelihood of following treatment schedules, and higher health care costs."[3]

Previously, most people with cancer didn't live long enough to develop chronic pain. Now, most do.

"As more cancer survivors look forward to full life spans, they face the increased possibility of chronic comorbidities and long-term treatment effects. Because these effects may not emerge until years after completion of treatment, the responsibility for detecting and managing them ultimately may lie with primary care physicians or other specialists.

"Additionally, as the number of long-term survivors continues to increase, the oncology workforce may not be sufficient to keep up with the needs of this population.

"'Oncology is a bit of a diminishing profession—we've got a whole group of excellent oncologists who are near retirement age, and not many new physicians are moving into the oncological realm,' Balazs Bodai, MD, director of the breast cancer survivorship institute at Kaiser Permanente Health, said in an interview with *Healio*. 'Oncologists who used to follow their patients forever can't follow them anymore because there are just too many people. So, the burden is going to fall on the primary care doctors to pick up the slack in the long-term management of these patients.'"[4]

Added Catherine M. Alfano, vice president of survivorship at the American Cancer Society, in the same *Healio* article: "Chronic effects of cancer therapy are those problems that were present during primary cancer treatment and continue in the posttreatment period. A late effect is something that was not present while treatment was going on, like during chemotherapy or radiation, but pops up seemingly out of nowhere a few years, even 10 years later.

"When it comes to who should be responsible for identifying and treating chronic and/or late effects of cancer treatments, Alfano said there is no simple answer. She noted that due to the current shortage of oncologists, PCPs frequently find themselves faced with the prospect of handling these cases. However, she said they generally are not trained to diagnose and manage the late effects of cancer treatments."[5]

"We think of pain as a symptom. But in those patients, the pain is the disease," said Clifford J. Woolf, director of the F. M. Kirby Neurobiology Center at Boston Children's Hospital and a neurology professor at Harvard Medical School.[6]

And sometimes, all the apparent facts can lead to a dead end.

Writes physician Leslie J. Crofford of Vanderbilt University: "Clinicians as people fear chronic pain, a symptom that demands attention and intrudes into every aspect of a person's life. Clinicians also loathe chronic pain, perhaps the symptom that brings more patients into our practices than any other but is also the symptom most likely to make us feel helpless as healers."[7]

This frustration extends to chronic pain sufferers, especially when they find themselves with few immediate options. If the pain becomes unbearable late at night or on a weekend, a visit to their regular physician is often out of the question. Moreover, that person has probably already been consulted about a condition that has not demonstrably changed since the last visit. This leaves small "doc in the box" clinics or the emergency room. Either way, chronic pain often becomes a square peg in the round hole of medical care.

According to the results of one survey of emergency department personnel: "The clinical challenge of treating patients in the ED stemmed from a mismatch between patients' needs and what the setting can deliver. Participants reported frustration with the system and with chronic pain patients' perceived inconsistencies and requirements. However, they also highlighted good practice and acknowledged their frustration around not being able to help this group.

"ED staff found people presenting at ED with chronic pain to be a challenging and frustrating population to treat. Staff was constrained by the fast-paced nature of their jobs as well as the need to prioritize emergency cases, and so were unable to spend the time needed by chronic pain patients. This was seen as being bad for staff, and for the patient experience. Staff suggested that care could be improved by appropriate information, signposting, and with time invested in communication with the patient."[8]

Yet time is an unbending tyrant, and it isn't just emergency room physicians who find themselves at its mercy. Given the recommended average consultation time of fifteen minutes per patient, a general practitioner or broad-based specialist can ill afford to throw this taut schedule into disarray by venturing down the rabbit hole of a chronic pain case.

Although undertaken for the benefit of that patient, such a quest inconveniences all the impatient others shifting their feet and reading old magazines in the waiting room. One overlong appointment can start the dominos falling, ensuring that every subsequent patient encounter will be more rushed and stressful for all concerned.

Then there is the problem of chronic pain outlasting the medications engineered to relieve it. Eventually, the body adjusts to the original dosage and requires more medication to keep the pain at bay. That's how the pain-ridden can become addicted, adding one problem to another.

The term *chronic pain* is in itself confusing, because it is hardly a monolithic condition. It might arise from an old injury that had supposedly healed, damaged nerves, or out of nowhere.

Perhaps the best-known of these conditions is fibromyalgia, which presents a variety of symptoms including increased sensitivity to touch or pressure, debilitating fatigue, bowel or bladder problems, numbness in the extremities, and bouts of depression. Based on some current research, it sometimes seems to exist in conjunction with sleep apnea or post-traumatic stress disorder. Or it could be genetic.

"The cause of fibromyalgia is unknown; however, it is believed to involve a combination of genetic and environmental factors.

The condition runs in families and many genes are believed to be involved. The pain appears to result from processes in the central nervous system and the condition is referred to as a 'central sensitization syndrome.'"[9]

Explains one patient: "A severe fibro flare . . . is like a horrific case of the flu. I ache all over. The flares come in cycles and I have become familiar enough with my body and the illness to usually know when one is coming on. My sleep, of course, is worse than normal. The more I hurt, the less I sleep, and the less I sleep, the more I hurt. And my arms feel like I have weights suspended from them. My term for my arms when they are like that is likened to the arms of a ragdoll. In addition to the tender points in the back and hip areas, my tender points tend to also be the shoulders, neck and chest."[10]

"Not everyone gets the same fibromyalgia treatment," notes physician and university professor Kevin Hackshaw. "We figure it out based on the symptoms that are most concerning for that patient. The key is that you need to find your 'magic' combination that works for you."

Or perhaps the patient in question is not afflicted with fibromyalgia, but reflex sympathetic dystrophy (RSD), also known as complex regional pain syndrome. "Like phantom limb syndrome, in which an amputee still feels pain or some other sensation where the amputated limb used to be, RSD might be described as the inappropriate aftermath to an injury of a nerve or soft tissue. This reaction can be far out of proportion to the seriousness of the wound and often continues long after the initial trauma seems to be healed."[11]

You can see from this crazy-quilt array of symptoms how difficult RSD can be to diagnose. To borrow from the old parable, doctors can become like blind men and the patient the elephant. Depending upon the emphasis, this condition can resemble fibromyalgia, postherpetic neuralgia, carpal tunnel syndrome, Lyme disease, and any of a dozen other ailments.

This post from a woman on the MDJunction chat room indicates the wide range of RSD symptoms.

"I've had RSD for 12 years but was just diagnosed a little over a year ago. I didn't have any injury that I know of, except for maybe a low back strain. Mine started out with a bang, affecting with pain from the waist down, inside and out, as well as gait problems, blood pressure problems, insomnia, migraines, and just disabling mind fog. I've been in and out of the hospital dozens of times for pain control, nausea, vomiting, etc. I had to have a colostomy, have a pain pump which I'm in the process of weaning off because it's never really helped anything and it's just one more thing to get off of."[12]

A few windows into RSD have been opened for physicians, however. Unlike some other mysterious and invisible nerve conditions that cause chronic pain, RSD often makes its presence obvious—not just to scans and X-rays, but to the naked eye. The skin can turn various colors (blue, black, red) and become shiny. The affected area can become cold to the touch. Tissues can swell. Bones can be compromised to the point at which they become obviously deformed.

If nothing else, this can be advantageous to a patient who is planning to petition for disability or is asking for compensation for a previous injury. According to the RSD Foundation, "Many patients who develop RSD/CRPS as the result of an injury do so within the context of legal liability. Some patients can be expected to defend their rights in courts of law. It is not uncommon for the defendant to accuse the patient of faking their condition, especially if there are no objective findings of RSD/CRPS documented on the medical record.

"Therefore, the attending physician must assess more than just subjective complaints (medical history). The physician must aggressively seek and document aggressive findings. For example, about 80 percent of RSD/CRPS cases have differences in temperature in opposite sides that may be colder or warmer. These temperature changes may be associated with changes in skin color. Furthermore, the temperature differences are not just static. The skin temperature can undergo dynamic changes in a relatively short period of time (within minutes).

"Thus, in that sense, the most effective aid in proving the existence of RSD/CRPS may be one of the oldest in the doctor's tool bag—the humble thermometer."[13]

An early diagnosis would also be helpful, but that can be achieved only after ruling out the myriad other conditions that RSD might mimic. By the time a patient has been through a series of general practitioners and down a number of diagnostic dead ends, the clock is already ticking.

Unlike RSD, which often seems to choose its victims at random, postherpetic neuralgia, or PHN, is directly tied to shingles, which is, in turn, an offshoot of common chicken pox. In fact, the two share the same herpes varicella zoster virus. In some cases, instead of fading away after causing chicken pox, this virus takes refuge in nerve tissue, lurking there for years. If it reappears, it's often ferocious—sometimes reemerging as shingles, sometimes worse.

The only good thing about shingles is that it, like chicken pox, generally goes away—unless the patient is elderly, has a weakened immune system, or is simply unlucky. That's when PHN sets in, a malevolent echo of shingles' damage to nerve endings. As with other chronic pain afflictions, the body doesn't seem to realize that the varicella zoster infection is gone.

Roughly one in five people with shingles later contracts PHN, mostly those older than fifty (more than a 50 percent chance). If you manage to reach the age of eighty, avoid chicken pox and shingles at all costs—in that group, PHN follows shingles in 80 percent of the cases.

In terms of symptoms, PHN is often described by those it torments as "chicken pox from hell." It burns, it aches, and it itches. The viral assault often leaves the skin highly sensitive to touch or pressure or even a breeze.[14]

Experts believe that shingles causes scar tissue to form next to the nerves, creating pressure. This causes the nerves to send inaccurate signals, many of the pain signals, to the brain. It is unclear why some patients go on to develop PHN.

As you can see, the world of chronic pain can be not only cloak-and-dagger but bait and switch, similar symptoms sliding in and out of a potential diagnosis. One of the most mystifying ailments to pin down is Lyme disease.

Most people know by now that Lyme disease is caused by the bite of certain types of ticks. Each spring, local media outlets throughout much of the country dutifully warn their listeners and readers of the danger with photos or drawings of ticks and perhaps an interview with a Lyme disease sufferer.

What makes Lyme disease so difficult to diagnose is its chameleon nature. So many of its symptoms mimic those of other illnesses, including arthritis-like joint pain, numbness in the hands and fingers, chills, fatigue, a rash, irregular heart rhythms, memory loss, and mood swings.

According to veterinarian Scott Taylor: "Lyme disease [LD] affects every tissue and every major organ system in the body. Clinically, it can appear as a chronic arthralgia [joint pain], fibromyalgia [fibrous connective tissue and muscle pain], chronic fatigue, immune dysfunction and as a neurological disease. LD may even be fatal in severe cases."[15]

Lyme disease differs from most chronic pain conditions in that it is caused by a bacteria passed along by a black-legged or deer tick. Like such conditions as shingles or RSD, however, its onset can occur long after the original encounter.

People who spend a lot of time outdoors often take tick bites for granted. We also get them through our pets. And when the symptoms of Lyme disease finally occur, so much time might have elapsed since the bite that it isn't considered as a possibility.

For Iowa resident Beth Feree, it wasn't until fourteen years later, after visiting more than a dozen doctors and receiving a bewildering variety of diagnoses, that she learned what was really behind her chronic pain. By then, the disease had invaded her muscles, joints, nerves, bones, and psyche.

"It's awful to have all these things happening and not know why," she said. "It got so bad that some of my bones spontaneously fractured. And I would have debilitating headaches and unbearable fatigue."[16]

Unfortunately, I get the feeling that playing sleuth has faded from our profession. In overworked, overbooked offices, "face time" is often kept to a minimum, and patients are sent on their way with a prescription or two. The hope is that the medication will work and the mystery will resolve itself without the need for further investigation.

My patients, by contrast, have been down that road. They have seen general practitioners, tried at least one drug, and been referred to me because their chronic pain has not responded to standard intervention. The one thing that all of these poor souls have in common is that they hurt. But why? And what can I do about it?

First, I ask all my patients to fill out a rather lengthy questionnaire prior to their initial visit. This tells me what sort of pain they have been experiencing, where it seems to be located, how long it has lasted, and perhaps something about its severity and cycles. I also want to know their medication history. Meanwhile, I try to get as much information as possible from the referring physician.

Some patients have been passed along to me rather quickly after it became obvious that the previous doctor was faced with something beyond his or her experience. Others have, in frustration, tried several different physicians and have accumulated a rather lengthy medical dossier.

Either way, my intent is to hit the ground running as a healthcare provider. With the patient's background in hand, I can spend more time asking detailed questions about what might be hurting them. At the same time, I try to minimize the expense to the patient by not duplicating tests that have already been done. Most of the referring physicians with whom I work are professional and thorough, and chances are they have already done MRI and CT scans and blood work. If so, there is no reason to go back over that road, because nothing will have changed in such a short time.

"Chronic pain is incredibly complex," said Benjamin Kligler, national director of the Integrative Health Coordinating Center at the Veterans Health Administration. "It is interwoven with all kinds of psychological, emotional and spiritual dimensions, as well as the physical. Honestly, the profession of medicine doesn't have a terribly good understanding, overall, of that kind of complexity."[17]

In 2021, this admitted lack of understanding led the National Institutes of Health to establish the Acute to Chronic Pain Signatures program (A2CPS) to "develop a set of objective biomarkers that provide 'signatures' to predict if chronic pain is likely to develop or be resolved after acute pain, like an injury or after a surgery. These signatures are greatly needed because prevention of chronic pain is a major challenge in pain management.

"This high prevalence of chronic pain has in part contributed to the current opioid epidemic in the US. A signature that could be identified before the transition from acute to chronic pain could help accelerate therapy development and ultimately guide pain prevention strategies."[18]

If nothing else, the opioid epidemic drew new attention to the chronic pain epidemic that was feeding it. The A2CPS program snapshot offers a clear indication of a shift in priorities.

"Many drugs, while effective early on, lose efficacy over time and make the transition from acute to chronic pain worse. In those who transition to chronic pain, maladaptive changes occur throughout the nervous system. Our ability to reverse these changes is very limited. Our lack of understanding of the mechanisms of transition to chronic pain is a major gap in knowledge that limits the development of effective preventive therapies. The ability to identify those at risk for transitioning to chronic pain could inform future clinical trials, improve the success of trials, and transform acute pain treatment approaches for prevention of chronic pain."[19]

This is, to be sure, a welcome change in attitude.

The genesis of chronic pain is complicated. Common knowledge tells us that people are hurt because something very specific has hurt them. The idea that the brain and the network of nerves connected to it can be mistaken seems somehow suspicious. Thus, many chronic pain sufferers face a double-edged sword—they are in pain and yet few believe that they are in pain. Sometimes, not even their physician.

A story on National Public Radio in 2019 featured a teenager named Devyn who described being suddenly accosted by fierce pain as mysterious as it was overwhelming.[20]

"It spread from my knee down to my entire foot, the whole bottom half of my leg, to my right leg, to my right arm. My entire arm, to my shoulder, to my left hand, and then my whole left arm," says Devyn. "Pain just took over. Sometimes I couldn't wear pants, I couldn't wear socks, I couldn't have the covers on my leg. Sometimes I'd have to turn the fan off, because the fan being on would hurt my leg."

Her mother, Sheila, at first feared the worst.

"You go to cancer first, right? It's like, 'OK, maybe you have cancer, maybe it's a tumor?' Sheila says."

When she was calm enough to reason with herself, Sheila decided cancer was improbable but wondered what was going on. The only thing they could think of was that the hip pain was somehow related to the minor knee surgery Devyn had a few months before—she had broken the tip of her distal femur one day during dance practice.

Sheila snapped to attention to solve the problem. It was 2016—surely modern medicine could fix this. She started by calling Devyn's surgeon, but the surgeon had no explanation for the pain. He renewed Devyn's prescription for Percocet and wrote a new prescription for Tramadol. But the pain worsened, so they lined up more appointments: Devyn's pediatrician, a naturopath, a pain specialist, a sports medicine doctor.

Every doctor's visit was the same. The doctor asked Devyn about her pain: where was it, and how severe was it on a scale one to ten? Then the doctor would order some tests to find the pain's cause.

But no matter where the doctors looked in Devyn, all they saw was a perfectly normal body. "You are healthy. Nothing is wrong."

Those are the words the doctors said to Devyn and Sheila over and over again. It made no sense. And it felt, paradoxically, like the more attention they gave to the pain, the bigger the pain grew.

"I remember one time when we were sitting in the doctor's office, and they basically just said to Devyn, 'You know you need to work really hard on your psychology' . . . and I thought, 'Maybe we're both crazy!' Sheila says."[21]

Individuals suffering from migraine headaches once received similar treatment—or lack of thereof.

"Nearly 30 million Americans suffer from migraines, a condition characterized by crushing pain, frequently on one side of the head, that is often coupled with nausea and vomiting, sensitivity to light and sound and sometimes even visual disturbances (known as auras)," wrote Laura Shocker in the *Huffington Post* in 2011. "A single attack can take anywhere from six to 48 hours to run its course. This very specific type of headache often runs in families and is typically brought on by a variety of triggers, which may include physical or emotional stress, changes in sleep patterns, certain odors and bright lights, among many others.

"But for a neurological condition as common as migraines, many people still doubt that it's a real condition. One recent study found that people with chronic migraines report feeling more rejected and ridiculed by friends, employers and even family members than patients with other types of neurological troubles, such as stroke, Parkinson's or Lou Gehrig's disease."[22]

In 1998, a Fresno State University psychology professor named Marcia Bedard presented a paper at the annual meeting of the Society for Disability Studies titled "Bankruptcies of the Heart: Secondary Losses from Disabling Chronic Pain." In it, she directly confronted the idea that most individuals claiming chronic pain did so seeking idleness, sympathy, or drugs.

"For more than 30 years now," Bedard wrote, "the majority of psychologists have been shifting their emphasis toward treating chronic pain as a perceptual and psychological phenomenon rather than a true medical problem. One of the major theorists in this field was Wilmer Fordyce, who developed an influential social-learning model of chronic pain based on behavioralism about 20 years ago. Fordyce belied that pain is behavior designed to protect oneself or solicit aid and that pain increases, i.e., this behavior is strengthened, when followed by desirable consequences." According to Bedard, Fordyce listed the four most common "secondary gains" as "1. Attention and sympathy from family, friends and physicians; 2. Release from task responsibilities at home and at work; 3. Narcotic medications are presumed to induce constant euphoria; and 4. Monetary compensation which approximates actual wages."

Bedard countered this with her own list of four "secondary losses": (1) Anger/trivialization/rejection by family, friends and physicians; (2) Complicated/frustrating tasks dealing with new bureaucracies; (3) Agonizing pain without medication; unpleasant side effects with medication; and (4) Denial of disability benefits to which they are legally entitled.

In her paper, Bedard made a telling argument that Fordyce's "gains" tended to be ephemeral, most often because of the longevity of chronic pain.[23]

Sympathy, it seems, has a shelf life. Whether it is a spouse or a family physician, human beings are generally oriented toward solving problems. A husband or wife gladly takes greater responsibility for the children or the housework to help a loved one recover from an illness or injury but understandably becomes frustrated and angry when that added burden drags on for months or years.

At the same time, it seems obvious in many cases that a person's chronic pain is affected by his or her mental state and level of anxiety. That's why I often recommend counseling along with the medical component.

"We said, 'Why don't we bring in some patients and look inside their brains?' said A. Vania Apkarian, a professor of physiology, anesthesiology and physical medicine and rehabilitation at Northwestern University's Feinberg School of Medicine. 'And as soon as we did that, we found all kinds of differences between healthy pain patients and chronic pain patients.'"[24]

There is no denying that the mind plays a key role in how pain is perceived. Studies have shown that a dental patient expecting a procedure to be extremely painful often invites that pain into his or her mouth. Similarly, with chronic pain patients, pain becomes so much a part of their lives—a path grooved so deeply into the brain—that it is accepted as that person's daily reality.

April Vallerand, a pain researcher and professor at Wayne State University in Detroit, says that a sense of powerlessness helps shape her patients' perceptions of pain.

"If you perceive yourself to be disabled, you're going to act like it," she says. "Patients would say to me, 'I'm fine as long as I don't move from that recliner.' Many were afraid to cook, drive, go to the mall. Well, that's not life, that's not function. My goal was to maintain or improve their function, despite chronic pain."

The key is restoring their sense of control, which is known to reduce pain-related emotional distress and improve function. Vallerand designed a program for cancer patients she called Power over Pain—Coaching (POP-C). It's delivered by trained nurses through phone calls and home visits. This establishes trust and helps caregivers understand patients' backgrounds, stories, and cultures—all essential for helping them learn to manage pain. POP-C has three main components: medication management, pain advocacy, and living with pain.[25]

Besides cancer patients, Vallerand also works with those suffering from various types of arthritis. Although generally regarded as one of the curses of growing older ("My arthritis is acting up again—must be fixing to rain"), it invades bodies of all ages. Though not life-threatening, it can be debilitating, especially when it invades the hands and knee joints.

"Osteoarthritis is the most common type of arthritis affecting 27 million Americans. Many people believe it's a crippling and inevitable part of growing old. But things are changing. Treatments are better, and plenty of people age well without much arthritis. If you have osteoarthritis, you can take steps to protect your joints, reduce discomfort, and improve mobility.

"Rheumatoid arthritis, on the other hand, is an inflammatory condition in which your immune system attacks the tissues in your joints. It causes pain and stiffness that worsen over several weeks or a few months. And joint pain isn't always the first sign of rheumatoid arthritis—sometimes it begins with 'flu-like' symptoms of fatigue, fever, weakness, and minor joint aches."[26]

Another condition sometimes grouped under the heading of chronic pain is one I deal with often in my practice—back pain.

According to a fact sheet from the National Institutes of Health, "The first attack of low back pain typically occurs between the ages of 30 and 50, and back pain becomes more common with advancing age. Loss of bone strength from osteoporosis can lead to fractures, and at the same time, muscle elasticity and tone decrease. The intervertebral discs begin to lose fluid and flexibility with age, which decreases their ability to cushion the vertebrae. The risk of spinal stenosis also increases with age.

"Back pain is more common among people who are not physically fit. Weak back and abdominal muscles may not properly support the spine. 'Weekend warriors'—people who go out and exercise a lot after being inactive all week—are more likely to suffer painful back injuries than people who make moderate physical activity a daily habit. Studies show that low-impact aerobic exercise can help maintain the integrity of intervertebral discs."[27]

Lots of things can go wrong with the back, from vertebrae to joints to muscles. Sometimes, the damage can be cumulative, as with someone who regularly moves heavy furniture or plays an "impact" contact sport. At other times, like many forms of chronic pain, its origin is a mystery.

Still, for most of medical history, back pain—albeit often difficult to treat—was regarded as a basic "meat and potatoes" affliction with few subtleties, something that attacked the bones or muscles but had little to do with the brain. Recent research, however, has added some new wrinkles.

"Brain imaging studies suggest that people with chronic low back pain have changes in the structure and function of certain brain regions. Other research seeks to determine the role of brain circuits important for emotional and motivational learning, and memory in this transition, in order to identify new preventive interventions. Furthermore, several studies are being conducted to identify and characterize bidirectional neural circuits that communicate between the spinal cord to brain, which are aimed at discovering and validating new interventional targets for low back pain."[28]

No matter what their specific area of interest might be, chronic pain researchers are accelerating the search for an elusive "off" switch. If the network of nerve fibers can turn on the body's reaction to pain, there should be a way to turn it off.

Scientists have discovered that certain chemicals—everything from globulin to potassium to histamine to various acids—trigger a response from the nociceptors. If the same chemicals are injected in very small doses, they generally cause immediate pain.

Are these chemicals, then, the physical manifestation of pain? Or are they only its enablers? And by what mechanism are they secreted when pain threatens?

According to the most commonly quoted count, one in three Americans suffers from some form of chronic pain. Given the nature of Western society, however, many of those affected are ill-equipped psychologically to deal with it. The myriad TV ads for pain relievers almost always imply a speedy relief from discomfort, a change immediately evident in the sufferer's body language and facial expression. The reality is that most pain medicines merely pull a temporary shroud over the hurt without dealing with the root cause.

As a society, we are not patient. When something hurts, we want and expect that unpleasant feeling to go away as quickly as possible. And when the normal cause and effect breaks down, it jolts our mental equilibrium.

Physicians often feel the backlash of this. We all hate to receive that phone call in which a strained voice on the other end tells us: "It *still* hurts. What do I do?"

Often, that's a tough question.

CHAPTER 7

Beware the Evil Twin

NOT SO LONG AGO, BEFORE EVERYTHING WENT TERRIBLY WRONG, chronic pain patients were a bonanza for the companies that produce opiates.

Some medications are prescribed only for the length of a short treatment—antibiotics, for example. Maintenance drugs (like those for lowering blood pressure) might be longer term, perhaps even over a lifetime, but are normally taken only once or twice a day.

Opiates were the best of both worlds from a pharmaceutical provider's point of view. The doses are generally more frequent, and the length of the prescription was often essentially open-ended.

In many ways, however, this was a recipe for disaster.

Opioids continue to be the flagship drug when it comes to defusing pain, both acute and chronic. Sometimes, they succeed in that function when nothing else does.

The problem is, the way in which they affect the human body can send their users down the path of diminishing returns. Although every individual might react differently to an opioid regime, most eventually begin to develop a resistance to it. When that happens, they need to take more pills to achieve the same effect, then more still. Lest we forget, heroin is an opioid.

On the surface, there is nothing easier for a doctor than writing a prescription—a couple of scribbles, a quick flip of the wrist to separate sheet from pad, an extended hand.

If the recipient is a chronic pain patient, however, this simple transaction can be a moment of profound anticipation. Chances are, many other medications have already failed, yet there is always hope that this will be the one to restore normal life.

For the physician, conversely, there is often a lurking sense of uneasiness. Perhaps gun shop owners—the responsible ones, anyway—have the same feeling when they hand over their merchandise.

Much can go right with today's drugs, but much can also go wrong. With first-time patients, I can only assume that they will follow my instructions and the information on the bottle and not take too much or too little. If the medication is an opioid, I trust that it will be used for pain relief and not recreation or the search for forgetfulness.

Increasingly, these are too many "ifs" for the average physician, which is why many of the stronger medications for chronic pain are now most often prescribed by pain specialists. Unfortunately, their ranks are dwindling instead of increasing.

These are some of the scenarios that could lead to misuse of an opioid drug:

1. An elderly person might take his or her medication several times in a short period because of a memory lapse.
2. A patient might misunderstand or misjudge the amount of time it takes a particular medication to "kick in." Thus, when no relief is felt in what that patient believes is a reasonable time, he or she may take another pill under the assumption that more is needed.
3. A depressed person, living alone, might be more inclined to combine too many pills with alcohol.

4. The same patient, in too much pain to work or even be active, might be tempted to get high on an extra pill out of boredom.[1]

No physician in his right mind wants his patients to overdose on the drugs he prescribes. Besides the emotional toll of contributing to a tragedy, such an event can result in a malpractice suit, serious damage to the doctor's reputation, or even law enforcement scrutiny and arrest.

Still, once the patient walks out of a doctor's office, his or her control over that person is severely limited. Yet just as it would be unfair to blame ten bartenders for a drunk-driving accident caused by a driver who stopped for only one drink at each bar, neither can a doctor be blamed if drugs obtained by prescription are misused or combined with something obtained through black-market transactions.

Wrote Sarah DeWeerdt in a 2019 *Nature* article: "Opioid addiction is not a new phenomenon in the United States, but in the past, it did not have such a marked impact on the nation as a whole. The groundwork for the crisis was laid in the 1980s when pain increasingly became recognized as a problem that required adequate treatment. US states began to pass intractable pain treatment acts, which removed the threat of prosecution for physicians who treated their patients' pain aggressively with controlled substances. And, in 1995, the American Pain Society, a physicians' organization in Chicago, Illinois, launched a campaign that framed pain as a 'fifth vital sign' that should be monitored and managed as a matter of course, in the same way as heart rate and blood pressure.

"Before the present epidemic, opioids were prescribed mainly for short-term uses such as pain relief after surgery or for people with advanced cancer or other terminal conditions. But in the United States, the idea that opioids might be safer and less addictive than was previously thought began to take root. A letter to the editor in the *New England Journal of Medicine* in 1980 reported that of 11,882 hospitalized people who were prescribed opioids, only four became addicted, but the short letter provided no evidence to back up these claims. A widely cited 1986 study, involving only 38 people, advocated using opioids to treat chronic pain unrelated to cancer. The prevailing view is that these studies were over-interpreted. But at the time, they contributed to the perception that opioids were addictive only when used recreationally—and not when used to treat pain."[2]

Although hospital patients are sometimes given powerful pain-relief drugs during a relatively short stay—whether by mouth, needle, or intravenously—the effects of those medications rarely last beyond the period of healing. There may be brief periods of euphoria, but not enough for the average person to fall in love with it.

Fortunately, for most people, taking strong anti-pain drugs in a hospital setting doesn't translate well into a normal life at home. It would be hard to function at a workplace or as a parent while in a constant state of euphoria.

Moreover, opioids are not "recreational drugs" in the usual sense of social communion. Besides limiting everyday function, they tend to pull the user into a soft cocoon, away from the outside world. Chances are, hardly anyone ever tells friends: "Why don't you come over tonight and we'll do some OxyContin while we watch the ballgame."

Still, there are unquestionably some individuals who try to incorporate these "happy drug" feelings into their daily existence. What tempts them are the endorphins that come as a side effect to pain relief.

The body also creates endorphins naturally—for example, the well-known "runner's high," or the "afterglow" following sex, both of which are described as part of an internal "reward system."

These are usually short-lived glimpses of euphoria, however, not nearly as powerful as those from artificial sources. In 2007, a group of medical researchers at the University of California–San Francisco found out why.

"It has generally been thought that all opioid molecules, natural or synthetic, impart their signal only from receptors on the surface of the cell. Opioid-bound receptors are then taken inside the cell to compartments called endosomes, but receptors were thought not to signal from this location. Overturning this long-held view, the research team discovered that receptors actually remain active in endosomes and they use the endosome to sustain the signal within cells.

"But in the most intriguing twist, the research team discovered that morphine and synthetic opioids activate receptors in yet another internal location called the Golgi apparatus, where endogenous opioids are unable to produce any activation at all.

"'It really surprised us that there was a separate location of activation for drugs in the Golgi apparatus that could not be accessed by endogenous opioids,' said first author Miriam Stoeber, a postdoctoral researcher. 'Drugs, which we generally thought of as mimics of endogenous opioids, actually produce different effects by activating receptors in a place that natural molecules cannot access.'

"Moreover, morphine and synthetic opioids crossed cell membranes without binding receptors or entering endosomes. They traveled directly to the Golgi apparatus, reaching their target much more quickly than endogenous opioids got into endosomes, taking only 20 seconds compared to over a minute. This time difference could be important in the development of addiction, the researchers said, because typically the faster a drug takes effect, the higher its addictive potential."[3]

In other words, these substances trump the body's natural reward system with something stronger, quicker, and of longer duration. The downside, however, is considerable. Once the body becomes accustomed to these new sensations, it soon wants more. Even worse, it often issues a strong protest when they are taken away.

Such stories too often end badly because increasing ingestion (or injection) of opioids is an overdose waiting to happen. It turns out that these normally useful drugs also harbor an evil twin that kills people.

"Preventable drug-related poisoning deaths, or drug overdoses, increased in 2019 after a slight decrease in 2018," reported the website *NSC Injury Facts* in 2020. "Preventable drug overdose deaths increased 5.5% in 2019, from 58,908 in 2018. In 2019, 62,172 people died from preventable drug overdoses—an increase of 457% since 1999. These deaths represent 88% of the total 70,630 drug overdose deaths in the United States, which also include suicide, homicide, and undetermined intents."[4]

According to a 2018 article in *Scientific American*: "When a person smokes, snorts or injects an opioid, the substance enters the bloodstream, then the brain. There it can act on mu-opioid receptors, says Eric Strain, director of the Center for Substance Abuse Treatment and Research at Johns Hopkins University.

"'Once the drug binds to those opioid receptors and activates them, it sets off a cascade of psychological and physical actions; it produces euphoric effects, but it also produces respiratory-depressing effects,' Strain says.

"As a result, victims of a fatal overdose usually die from respiratory depression—literally choking to death because they cannot get enough oxygen to feed the demands of the brain and other organ systems. This happens for several reasons, says Bertha Madras, a professor of psychobiology at McLean Hospital and Harvard Medical School. When the drug binds to the mu-opioid receptors it can have a sedating effect, which suppresses the brain activity that controls breathing rate. It also hampers signals to the diaphragm, which otherwise moves to expand or contract the lungs. Opioids additionally depress the brain's ability to monitor and respond to carbon dioxide when it builds up to dangerous levels in the blood.

"'It's just the most diabolical way to die, because all the reflexes you have to rescue yourself have been suppressed by the opioid,' Madras says."[5]

Still, those who continue to abuse these drugs see some definite cost advantages over illegal substances like heroin (also an opioid, but in a different legal category), methamphetamines, and even marijuana. Because opioids are legal, they could be obtained using a doctor's prescription—or from an acquaintance who had managed to amass an extra supply. Except for the most desperate, it wasn't necessary to risk entering the dark realm of street-level dealers. And wasn't this stuff just medicine, after all?

Indeed, many—if not most—opioid addicts don't fit the negative drug-user stereotype. These are people who entered the world of opioids simply to seek relief from pain. So who's fault is this crisis? Wrote physician Ronald Hirsch in the *Journal of the Missouri State Medical Association*:

"First, in all fairness, I will start with physicians. We overprescribe opioids, just as we overprescribe antibiotics. But it is generally well-meaning; we don't want our patients to experience pain. But then we prescribe 30 or 60 pills when 5 or 20 would have been adequate. We do that because we are used to prescribing in multiples of 30; 30 days for a month supply of a once-a-day medication, 90 days for a mail-order prescription. Prescribing 6 or 10 pills will undoubtedly result in a phone call from a pharmacist asking for a round number of pills, taking up time better spent entering meaningless information into our electronic health record systems.

"It is the leftover pills that sit forgotten in the medicine cabinet which often lead to trouble, stolen by a relative or visitor and abused. But sometimes it is that prescription that was provided for true pain that leads rapidly to tolerance and addiction. *Healthy Living* magazine recently published a heart-wrenching story of a woman whose life was nearly destroyed by two weeks of oxycodone prescribed by a well-meaning physician for arthritis.

"The role of these physicians can best be described as innocent bystander. We were truly trying to help the patient. But there are also what are known as 'pill mill' doctors who set up shop, accept cash as the only payment and are willing to prescribe to anyone for any ailment, real or feigned. One physician in my area was so bold as to meet his 'patients' in a local coffee shop to exchange prescriptions for cash. Needless to say he is no longer licensed to practice medicine. Doctors such as these are criminals and need to be stopped. They cast a long shadow on the work of every other physician trying to help patients."[6]

Hirsch went on to level harsh criticism at a number of major pharmaceutical companies that he said were aided and abetted by several inattentive government agencies.

As the NIH's National Institute on Drug Abuse reported in 2018: "In the late 1990s, pharmaceutical companies reassured the medical community that patients would not become addicted to prescription opioid pain relievers, and healthcare providers began to prescribe them at greater rates. This subsequently led to widespread diversion and misuse of these medications before it became clear that these medications could indeed be highly addictive."[7]

Perhaps not coincidentally, it was in 1995 that Connecticut-based Purdue Pharma came out with OxyContin.[8]

"Prescriptions for opioids increased gradually throughout the 1980s and early 1990s. But it wasn't until the mid-1990s, when pharmaceutical companies introduced new opioid-based products—and, in particular, OxyContin, a sustained-release formulation of a decades-old medication called oxycodone, manufactured by Purdue Pharma in Stamford, Connecticut—that such prescriptions surged and the use of opioids to treat chronic pain became widespread.

"Purdue Pharma and other companies promoted their opioid products heavily. They lobbied lawmakers, sponsored continuing medical-education courses, funded professional and patient organizations and sent representatives to visit individual doctors. During all of these activities, they emphasized the safety, efficacy and low potential for addiction of prescription opioids."[9]

Medical products, especially prescription-only medications, are something of an anomaly in the advertising world. Instead of targeting consumers, the drug companies have always focused on the physicians who prescribe and control their product. It's rare that a patient will request a particular drug based on exposure to an ad—generally, the doctor takes the lead there, and the patients follow his or her suggestion.

Direct kickbacks from pharmaceutical companies to physicians are banned by a US statute, so firms like Purdue enticed and rewarded "their" doctors by inviting them to what was advertised as seminars. Of course, these seminars were much more likely to be held in Phoenix, San Diego, or some other attractive destination than in Cleveland or Detroit.

"From 1996 to 2001, Purdue conducted more than 40 national pain-management and speaker-training conferences at resorts in Florida, Arizona, and California. More than 5,000 physicians, pharmacists, and nurses attended these all-expenses-paid symposia, where they were recruited and trained for Purdue's national speaker bureau. It is well documented that this type of pharmaceutical company symposium influences physicians' prescribing, even though the physicians who attend such symposia believe that such enticements do not alter their prescribing patterns.

"One of the cornerstones of Purdue's marketing plan was the use of sophisticated marketing data to influence physicians' prescribing. Drug companies compile prescriber profiles on individual physicians—detailing the prescribing patterns of physicians nationwide—in an effort to influence doctors' prescribing habits. Through these profiles, a drug company can identify the highest and lowest prescribers of particular drugs in a single zip code, county, state, or the entire country. One of the critical foundations of Purdue's marketing plan for OxyContin was to target the physicians who were the highest prescribers for opioids across the country. The resulting database would help identify physicians with large numbers of chronic-pain patients. Unfortunately, this same database would also identify which physicians were simply the most frequent prescribers of opioids and, in some cases, the least discriminate prescribers."[10]

Of course, Purdue was hardly the only guilty party here—just the most visible and blatant. In 2014 and 2015, opioid manufacturers paid hundreds of doctors across the country six-figure sums for speaking, consulting, and other services. Thousands of other doctors were paid more than $25,000 during that time. Physicians who prescribed particularly large amounts of the drugs were the most likely to get paid.

"'This is the first time we've seen this, and it's really important,' said Dr. Andrew Kolodny, a senior scientist at the Institute for Behavioral Health at the Heller School for Social Policy and Management at Brandeis University, where he is co-director of the Opioid Policy Research Collaborative.

"'It smells like doctors being bribed to sell narcotics, and that's very disturbing,' said Kolodny, who is also the executive director of Physicians for Responsible Opioid Prescribing.[11]

"'I don't know if the money is causing the prescribing or the prescribing led to the money, but in either case, it's potentially a vicious cycle. It's cementing the idea for these physicians that prescribing this many opioids is creating value,' said Dr. Michael Barnett, assistant professor of health policy and management at the Harvard T. H. Chan School of Public Health."[12]

Once the opioid crisis was finally identified as a serious problem, several major pharmaceutical companies were assessed heavy fines, and a number of physicians were sent to prison. Yet the wheels of justice hardly revolved swiftly.

In 2007, the federal government brought criminal charges against Purdue Pharma for misleadingly advertising OxyContin as safer and less addictive than other opioids. The company and three executives were charged with "misleading and defrauding physicians and consumers."[13]

"Purdue Pharma and the executives pleaded guilty, agreeing to pay $634.50 billion in criminal and legal fines. The three executives pleaded guilty on criminal misdemeanor charges."[14]

Nevertheless, it wasn't until 2019 that the case was finally resolved, with Purdue agreeing to file bankruptcy and pay $10 billion to settle a host of civil claims. The three charged executives escaped with probation.

Another year then dragged on before Purdue paid up entirely.

"The abuse and diversion of prescription opioids has contributed to a national tragedy of addiction and deaths, in addition to those caused by illicit street opioids," said Deputy Attorney General Jeffrey A. Rosen. "With criminal guilty pleas, a federal settlement of more than $8 billion, and the dissolution of a company and repurposing its assets entirely for the public's benefit, the resolution in today's announcement re-affirms that the Department of Justice will not relent in its multi-pronged efforts to combat the opioids crisis."[15]

Five years earlier, the Drug Enforcement Administration (DEA) announced that it had arrested 280 people, including twenty-two doctors and pharmacists, after a fifteen-month sting operation centered on health-care providers who dispense large amounts of opioids. The sting was dubbed "Operation Pilluted."[16]

Belated as the eventual outcomes may have been, these cases and the mind-boggling fines that emerged from them largely derailed the opioid gravy train for pharmaceutical companies, rogue physicians, and others in that pipeline. By the time of the Purdue settlement, however, it was already too late to stem the tide. Millions of pills had found their way into the black market, and opioid-related deaths kept climbing.

Even worse, much of the supply had passed from American business owners like Purdue's Sackler family to the even more ruthless Mexican cartels.

"Cartels found ways to get drugs into the U.S. despite COVID-related travel restrictions. More than 83,000 people died from drug overdoses in the 12-month period ending in July of 2020—the most overdose deaths ever recorded, according to data released this year by the U.S. Centers for Disease Control and Prevention. That's 18% of the 12-month period ending July 2019.

"Fentanyl, a man-made opioid, remains a top killer in the United States, according to the Drug Enforcement Administration's annual National Drug Threat Assessment. Users and even many dealers don't know that fentanyl may be hidden inside pills, dyed blue and stamped to look like the prescription pain pill oxycodone or OxyContin. On the streets, the pills are known as 'Mexican Oxy' or 'M30s' because the cartels stamp one side with '30' and the other with an 'M.'"[17]

Meanwhile, beyond the wide reach of the cartels were hundreds of thousands of Americans who were already hooked by the torrent of opioids dispensed by physicians on the take. In many cases, opioids became a gateway drug for even more problematic illegal substances.

"Pooling data from 2002 to 2012, the incidence of heroin initiation was 19 times higher among those who reported prior nonmedical pain reliever use than among those who did not (0.39 vs. 0.02 percent). A study of young, urban injection drug users interviewed in 2008 and 2009 found that 86 percent had used opioid pain relievers nonmedically prior to using heroin, and their initiation into nonmedical use was characterized by three main sources of opioids: family, friends, or personal prescriptions."[18]

Meanwhile, a study of opiate addiction in eastern India found that "persons dependent on opioids presented earlier for treatment, with earlier development of withdrawal symptoms and having completed lesser years of formal education, and had higher monthly incomes as compared to those dependent on alcohol. The most common psychosocial factors determining initiation and maintenance were peer pressure or curiosity."[19]

Once the line between legal and illegal drug use has been crossed with opioids, however, the temptation often extends to stimulant drugs such as cocaine and methamphetamines. Even opioids themselves can be lethal when obtained from unfamiliar sources, especially if fentanyl (fifty times stronger than morphine) has been stirred into the mix.

One high-profile overdose death involved Los Angeles Angels pitcher Tyler Skaggs, who was found dead in his Dallas hotel room in July 2019. An autopsy showed that Skaggs had significant amounts of oxycodone, fentanyl, and grain alcohol in his blood—an unimaginably lethal combination. Yet his death was ruled accidental, not a suicide.

The person charged with supplying Skaggs with his final high was the public relations director of his baseball team.

Sometimes, all the drug cops in the world can't protect us from ourselves. That's why the drug naloxone has often stood between an overdose scenario and death.

"Naloxone is a medicine that rapidly reverses an opioid overdose. It is an opioid antagonist. This means that it attaches to opioid receptors and reverses and blocks the effects of other opioids. Naloxone can quickly restore normal breathing to a person if their breathing has slowed or stopped because of an opioid overdose. But, naloxone has no effect on someone who does not have opioids in their system, and it is not a treatment for opioid use disorder."[20]

The way it has been described almost sounds like a game of tic-tac-toe—a synthetic opioid neutralizes the natural opioid, naloxone neutralizes the synthetic opioid.

"Naloxone is being used more by police officers, emergency medical technicians, and non-emergency first responders than before. In most states, people who are at risk or who know someone at risk for an opioid overdose can be trained on how to give naloxone. Families can ask their pharmacists or health care provider how to use the devices."[21]

Nevertheless, though it is unquestionably helpful—and, at times, lifesaving—naloxone has its limits.

"Naloxone works to reverse opioid overdose in the body for only 30 to 90 minutes. But many opioids remain in the body longer than that. Because of this, it is possible for a person to still experience the effects of an overdose after a dose of naloxone wears off. Also, some opioids are stronger and might require multiple doses of naloxone. Therefore, one of the most important steps to take is to call 911 so the individual can receive immediate medical attention. NIDA is supporting research for stronger formulations for use with potent opioids."[22]

As Mark Tyndall and Zoe Dodd pointed out in a 2020 article for the *AMA Journal of Ethics*, what is usually labeled "the opioid crisis" is actually two crises in one, each segment requiring a different approach.

It was easier to go after greedy drug companies and irresponsible physicians, because there was often a clear paper trail leading investigators straight to the source of abuse. Those who were injured by this were most often simply individuals trying to rid themselves of constant pain.

On the other hand, strictly "recreational" opioid use proved just as hard to stop as the illegal networks supplying heroin, methamphetamines, or cocaine. As any drug cop will tell you, the hardest crimes to solve and punish are those in which one party is offering something that the other party desires, minus the easy-to-track trail of recorded transactions.

Wrote Tyndall and Dodd: "In the United States, overdose deaths contribute more to a reduction in life expectancy than chronic lower respiratory diseases, Alzheimer's, or flu, and, in British Columbia, Canada, overdose deaths contribute to a decrease in life expectancy among those of lower socioeconomic status.

Although much of the media focus has been on prescription opioids, the majority of overdoses result from illicit drugs containing synthetic opioids with unpredictable potency. People buying these drugs run the constant risk of using toxic drugs and overdosing.

"The primary narrative that has emerged in the media and that dominates the public discourse has been to target those most affected including the people who use drugs, the communities that have been hardest hit by the crisis, and, more recently, the pharmaceutical companies that manufacture prescription opioids. However, the criminal policies that support the war on drugs have not changed even in response to this unprecedented crisis. The criminalization of users of illicit drugs has led directly not only to users' incarceration, but also to their marginalization and isolation and to violence, entrenched poverty, and a vicious cycle of trauma.

"At all levels, and by any measure, the response to such a massive and ongoing loss of life has been inadequate, as it has focused on prescribing and its downstream effects. Nearly all 50 states have prescription drug monitoring programs (i.e, databases that track controlled substance prescriptions) that provide health or law enforcement authorities with access to clinical data on prescribing patterns. The Centers for Disease Control and Prevention has focused specifically on monitoring trends, enhancing data collection, partnering with health systems to treat addiction and with community organizations and first responders to prevent overdoses and increasing public awareness about opioid use risk.

"In British Columbia, Canada, a public health emergency was declared in April 2016 that resulted in expanding treatment options, scaling up naloxone programs, and opening up new safe injection sites across the province. Although these initiatives have saved lives, there has been little movement on drug policy reform or on dealing with the contaminated illicit drug supply.

"The failure to act more decisively can only be explained by the entrenched discrimination and stigma against people who use drugs.

"In any other epidemic, such as an infectious outbreak, we would not even consider criminal enforcement as a response. Saving lives would be the priority. Our first response should be to provide a non-toxic, regulated alternative. In the case of the illegal opioid market, it is clear that removing the toxic product is just not possible, which should leave no alternative but to provide safer options in the form of a regulated opioid program."[23]

No one is asking personal care physicians or pain specialists to moonlight as DEA agents. Still, as Dr. Michael Schatman, editor of *Personal Pain Management*, points out:

"While the physician is morally obligated to relieve pain and suffering, he/she must simultaneously attempt—within his or her practice milieu—to identify potential abusers and diverters. Unfortunately, the 'science' of risk assessment is still in its infancy, and selection remains an overwhelming challenge. Ideally, physicians prescribing opioids to patients with chronic pain will consult with an appropriately trained and experienced pain psychologist in order to obtain an evaluation that will serve as a guide in the selection process. Tragically, however, access to qualified pain psychologists appears to be becoming progressively more limited, with this scarcity more pronounced in non-urban areas."[24]

A database of individuals charged with drug-related crimes would be helpful in this regard. As is a frequently shared list of red flags that might apply to new patients.

That, however, assumes that only individuals with a history of substance abuse are likely to abuse opioids or other pain-relief drugs. Chronic pain has a way of altering personalities and circumstances, putting some people on a dangerous path who would never have chosen it before.

"While the body of published research addressing risk factors for opioid abuse in general is substantial, investigations identifying chronic pain patients at risk for prescription opioid abuse are relatively scarce. It is important to recognize that neither the risk factors identified in the general population nor the assessment tools utilized to identify can necessarily be generalized to the chronic pain population, as the circumstances of their distress and predisposing psychological factors of the groups are not identical."[25]

As a physician specializing in chronic pain, I can't follow my patients home to make sure they are taking the drugs I prescribe responsibly. Nor do I have the authority (or the desire) to go in search of illicit opioid dealers.

Nevertheless, the medical community is left as the first line of defense in "accidental abuse." We need to make sure none of those medications are overprescribed. We must avoid allowing patients to receive more pills when their schedules show they should still have some left. We need to make sure they clearly understand how these opioids will affect them and the price they might pay for exceeding the recommended dosage.

Most of all, we need to keep looking for ways to curb our patients' pain without putting them at risk.

Chapter 8

Anomalies

Rules are made to be broken, we're told, and that's what this chapter is all about.

Feeling No Pain

We've learned that pain is universal, an inevitable part of every human existence. And so it is—for the overwhelming majority. However, there is also a tiny sliver of the human species for whom pain literally does not exist. The condition is called congenital insensitivity to pain (CIP for short), and it turns out to be more of a curse than a blessing.

Most of us who do feel pain will go to great lengths to avoid it, won't we? Well, yes, except for those who actually find it pleasurable.

Let's talk about this.

Wouldn't it be wonderful to be completely free from pain, all the time?

Be careful what you wish for.

"People assume that feeling no pain is this incredible thing and it almost makes you superhuman," Stefan Betz told the website *BBC Future*. "For people with CIP it's the exact opposite. We would love to know what pain means and what it feels like to be in pain. Without it, your life is full of challenges."[1]

According to the *BBC* site, "As a young child, Betz's parents initially believed he was mildly mentally retarded. 'We couldn't understand why he was so clumsy,' his father Dominic remembers. 'He was constantly bumping into things and getting all these bruises and cuts.'

"Neither his parents or siblings have the condition, but the diagnosis of CIP eventually came when at age five, he bit off the tip of his tongue, without any apparent pain response. Shortly afterwards he fractured the right metatarsal in his foot, after jumping down a flight of stairs."[2]

What Betz and others in his situation have discovered—often the hard way—is that CIP is not a superpower. Moreover, the fact that they are not in pain doesn't mean they aren't injured.

As discussed earlier, pain is primarily a protector rather than a villain. The brain uses pain to warn us against something that is unhealthy or risky, and it demands that we stop engaging in any activity that hurts us. Those with CIP, however, are on a different frequency.

In 2017, a newspaper article detailed the sad case of Indiana resident Amanda Smith.

"'When you can't feel pain, you never know when you're hurt,' she said. 'Not until it's too late.'

"Amanda's condition is slowly killing her, taking her body piece by piece.

"It started when she was 12. She scratched her toe on the floor of a public pool, and the infection spread to her blood before anyone knew she was hurt. Her right leg went next, followed by more of her left, and so on. Now 25, her left arm is her only remaining limb. She may lose that, too.

"'Hopefully that won't come for a long time, because I'm left-handed, so—' Amanda's voice trailed off.

"'I just don't want to lose anything else.'"[3]

One of the reasons scientists believe CIP is so rare is because so few individuals with the disorder reach adulthood.

"We fear pain, but in developmental terms from being a child to being a young adult, pain is incredibly important to the process of learning how to modulate your physical activity without doing damage to your bodies, and in determining how much risk you take," explained Dr. Ingo Kurth, Stefan Betz's physician in Germany.[4]

A broken leg is a broken leg, with or without pain. And medical emergencies such as appendicitis and aneurysms usually announce their arrival with acute discomfort that is lost on someone with CIP.

This condition was first discovered in 1932 by New York physician George Dearborn, who found it in one of his patients. Despite its quirkiness, however, CIP spent the next seventy years or so hiding on the back pages of medical journals and being discussed primarily anecdotally among doctors who dealt with the pain.[5]

That began to change with the arrival of social media and the appearance of CIP support groups worldwide. Suddenly, "scientists began to realize that studying this rare disorder may provide a new understanding of pain itself, and how to switch it off for the many afflicted by chronic pain conditions."[6]

Researchers have already discovered that CIP is caused by mutations in the NTRK1 gene, which gives the body instructions to make a protein that is important for the development and survival of nerve cells—especially those that carry information about pain, temperature, and touch (sensory neurons). As with many other genetic conditions, however, the dilemma is what to do about it.

Another version of CIP is coupled with anhidrosis, the inability to sweat, and known as CIPA. It is extremely dangerous in hot weather because the body's normal cooling mechanism has been switched off and death from heatstroke can happen quickly.

Fortunately, no one can "catch" either version—as genetic glitches, CIP and its variants are always present from birth.

According to the *American Journal of Medical Genetics* research, "it is especially common to those of Israeli Bedouin backgrounds, who make up half of the known cases. This prevalence is probably due to a high rate of consanguineous marriage."[7]

Congenital insensitivity to pain is also found in Vittangi, a village in Kiruna Municipality in northern Sweden, where nearly forty cases have been reported. The south Indian state of Telangana also has a relatively high number of patients with CIP.[8]

As Stefan Betz's parents discovered, having a child with CIP can add considerable stress to the already difficult job of parenting.

"Teething is a big hurdle to overcome for children with congenital insensitivity to pain. Just like any teething baby, they want to gnaw on everything in sight. But because they don't feel pain, they don't know when to stop. They might chew through their tongues or bite their fingers until they bleed. Babies don't understand directions, so you can't just tell them not to bite so hard. Some parents find it makes sense to remove the child's teeth—once the child's adult teeth grow in, he'll be old enough to understand when not to bite. But this removal of teeth can make eating more difficult. At a time when other children their age are learning to enjoy solid food, kids with CIPA are developmentally behind. Also, people with CIPA often don't feel hunger pains, so eating feels like an unnecessary chore.

"Other very common injuries for children with CIPA are corneal abrasions and other serious eye injuries. These come from scratching or rubbing the eyes too hard. Some children have to wear protective eyewear or use special eye medicines."[9]

Another issue is that the appearance of multiple bruises and other injuries on a child may cause observers to suspect child abuse.[10]

There is no cure for CIP, at least so far. Indeed, CIP patients seem most useful to the medical profession at this point as possible sources of new information.

"Future research with CIP patients may identify other proteins specifically involved in nociception, which might represent potential targets for chronic pain treatment," explains one research paper. "Moreover, this rare clinical syndrome offers the opportunity to address interesting neuropsychological issues, such as the role of pain experience in the construction of body image and in the empathic representation of others' pain."[11]

Another intriguing piece of the puzzle was the discovery that people with CID also had lost their sense of smell (as do many who contract COVID-19).

"To give birth painlessly is every pregnant woman's dream, but for people who lack the ability to feel pain, life stinks. Actually, they wouldn't know, because people with congenital insensitivity to pain can't smell anything.

"Researchers have discovered that these individuals who have a rare genetic condition rendering their lives pain-free also don't have a sense of smell. And they think they know why: Humans use the same signaling channel in the brain to both feel pain and sense smells."[12]

"Study researcher John Wood, of the University College London, tested the smelling abilities of three patients unable to feel pain. These patients have a mutation in an ion channel in their sensory cells called Nav1.7 that sends pain signals from the skin to the brain. That's when he found that this channel is also active in the olfactory system, and these patients had no sense of smell.

"'The results in the humans were very clear. They can't smell,' co-researcher Frank Zufall of the University of Saarland School of Medicine in Germany told LiveScience. 'It was completely surprising and completely unexpected.'

"A couple of previous studies of these pain-insensitive patients have hinted that they may not be able to smell, but this was the first to actually test their abilities. These three participants couldn't identify any smells (like balsamic vinegar, orange, mint, and coffee) that the researchers threw at them, even though they've lived relatively normal lives.

"To study how this Nav1.7 might also be linked with smell, Zufall and his colleagues examined the cells of the human olfactory system, and also studied genetically engineered mice without this channel in their smell systems. He found that in both humans and mice the channel works as an interpreter between the smell-sensing cells in the nose and the smell-interpreting cells in the brain."[13]

Most conditions this rare (literally fewer than one in a million people have it) tend to fly under the radar of public interest. In this case, however, the combination of its quirkiness and its negation of a nearly universal human quality (the ability to feel pain) has gradually eased CIP into the spotlight.

Ashlyn Blocker, a teenage girl from Florida with CIP, has been interviewed on *Good Morning America*. Plus, "an episode of the television medical drama *House* featured 16-year-old Hannah and her mother, who were in a severe automobile accident. Hannah saw her mother unconscious and dialed 911. She looked down and noticed a metal rod jammed in her own thigh. At the hospital, Hannah soon ran a 105-degree fever, although she neither sweated nor shivered. She then collapsed, and ice water and cooling packs were applied."[14]

In the early 2000s, a small Canadian biotech company called Xenon Pharmaceuticals heard about a family from Newfoundland in which several members of the family were affected by CIP. "The boys in the family had often broken their legs and one even stood on a nail without any apparent sense of pain," says Simon Pimstone, president and CEO of Xenon.

"The company began scouring the globe for similar cases, to try and sequence their DNA. The resulting study found a common mutation in a gene called SCNP9A, which regulates a pathway in the body called the Nav1.7 sodium channel. The mutation knocked out this channel, and with it, the ability to feel pain.

"It was the breakthrough the pharmaceutical industry had been waiting for.

"'Drugs which inhibit the Nav1.7 channel could be a new way of treating chronic syndromes such as inflammatory pain, neuropathic pain, lower back pain and osteoarthritis,' says Robin Sherrington, senior vice-president of business and corporate development at Xenon, who was heavily involved in the initial study. And because all sensory functions remain normal in CIP patients apart from the lack of pain, it offers the prospect of minimal side effects.

"Over the past decade, Nav1.7 has sparked a 'pain race' across the biotech industry between pharmaceutical giants including Merck, Amgen, Lilly, Vertex, and Biogen, all vying to become the first to bring an entirely new class of painkiller to market.

"But developing sodium channel blockers which act specifically on the peripheral nervous system isn't entirely straightforward, and while the promise is there, it may take another five years to fully know whether inhibiting Nav1.7 is really the key to modulating pain signaling in humans. Xenon themselves are banking that it is. They currently have three products in clinical trials in partnership with Teva and Genentech, one in phase two trials for shingles pain, and two more in the first phase of safety studies."[15]

Another drug being tested for CIP is naloxone, best known as a quick-acting antidote for opioid overdoses.

"It aims to counteract severe, life-threatening depression of the nervous system. Opioids block the sensation of pain and naloxone stops this blockage. This treatment only works in some CIPA patients and its mechanisms are not entirely known.

"One of the harsher, but effective treatments of CIPA is to remove all of the patient's baby teeth when they first come in. Sometimes children with CIPA can completely chew their tongues off for lack of feeling. This can cause the child to bleed to death if the behavior or disease is not diagnosed. This removal saves the fingers, lips, and tongues of children who are unable to stop before damage is done. By the time the patient's adult teeth begin to grow in, he or she will likely have developed an understanding of healthy behavior.

"Most of the studies on CIPA have focused more on its mechanism in development more than on creating possible treatments for the disorder. Dr. Felicia Axelrod, a pediatric neurologist at New York University who studies CIPA, highlighted the importance of this research. 'Understanding mechanisms of autonomic control and pain perception in CIPA . . . is very special, as these are genetic models that help us understand function and maintenance of the autonomic and sensory nervous systems,' she said. Scientists have begun to identify pathways absent in CIPA patients that may be essential for developing the entire nervous system, not just pain neurons.

"'Right now we are still trying to understand the mechanisms, but the goal of course will be how to use this knowledge to help the patients,' Dr. Axelrod said."[16]

CIP and CIPA are not potential moneymakers for the pharmaceutical industry, of course—not given their tiny slice of the potential market. That's a plus for naloxone, which is well-known, has already been developed, and could easily serve another purpose.

Meanwhile, continuing research into the relationship between the brain and the nociceptors keeps popping up in a wide variety of medical quests.

"The past two decades have seen unprecedented advancement in understanding the genetic complexity of nociception and in the diagnosis of human genetic disorders and their underlying molecular pathology. These discoveries have ushered a number of clinical trials to test the efficacy of the next generation of pain therapeutics. The complications and adverse side effects have plagued these efforts, underscoring the complexity of nociceptive biology. Monogenetic pain disorders have been instrumental in shedding light on this complexity; however, we are only at the beginning of this journey."[17]

For many of those afflicted by CIP and CIPA, however, a drawn-out research time line is a luxury they can ill afford.

"While the world of painkiller research is benefiting from the uniqueness of those with this extraordinary disorder, for CIP sufferers themselves, the prospect of a future life with pain and all its advantages remains uncertain."[18]

At least they are no longer alone and ignored.

When It Hurts So Good

Thus far, we've talked about the causes of pain, how to stop it, and even what it's like not to feel it. But what about those among us who actually seem to enjoy pain and seek it out? How does someone become a masochist?

According to a 2015 article on the website *BBC Future*: "The link between pleasure and pain is deeply rooted in our biology. For a start, all pain causes the central nervous system to release endorphins—proteins which act to block pain and work in a similar way to opiates such as morphine to induce feelings of euphoria.

"The relationship will come as no surprise to those who run."[19]

Still, "runner's high" is generally regarded as a side effect, not the primary incentive. But then there's Jason McNabb, who competes in eating contests involving bhut jolokia peppers that are two hundred to four hundred times hotter than jalapeños. No one forces him to do this, but for McNabb, "the pain from the peppers produces a rush that is similar to that produced by food, drugs or sex.

"'The pain subsided pretty quickly and then it was just the high of the adrenaline and euphoria from the peppers,' Jason explained."[20]

McNabb is also rewarded for his feats of endurance—victory in a contest and the monetary prize that comes with it. Yet the article about his exploits didn't say whether he sometimes devours those ferocious peppers at home, just for fun.

Indeed, many people endure pain as part of some activity that they view as constructive and ultimately reward bearing. You may have heard the old joke about the man who is asked why he keeps hitting himself in the head with a hammer.

"Because it feels so good when I stop," he explains.

For most of us, though, the popular concept of masochism evokes images not of hot pepper eaters or long-distance runners, but sexual adventurism.

Encyclopedia Britannica defines the word rather narrowly as "a psychosexual disorder in which erotic release is achieved through having pain inflicted on oneself. The term derives from the name of Chevalier Leopold von Sacher-Masoch, an Austrian who wrote extensively about the satisfaction he gained by being beaten and subjugated. The amount of pain involved can vary from ritual humiliation with little violence to severe whipping or beating; generally the masochist retains some control over the situation and will end the abusive behavior before becoming seriously injured. While pain may cause a certain amount of sexual excitement in many persons, for the masochist it becomes the chief end of sexual activity."

And yet, that entry adds, "The term is frequently used in a looser social context in which masochism is defined as the behavior of one who seeks out and enjoys situations of humiliation or abuse."[21]

We hear that all the time.

"I can't believe I'm working overtime again tonight. I must be a masochist."

Probably not, unless that overtime work involves whips and chains.

Sexual activity remains one of the frontiers of brain research. We still don't know why some of us are physically attracted to the same gender despite negating the biological impulse toward reproduction. Or why some people feel trapped in the wrong gender and opt for surgery to switch over. Masochism is just another puzzle to add to the list.

With those who seem to enjoy having pain inflicted on them, is that a brain function or simply the sign of a deep inferiority complex? David J. Linden addressed that issue in a 2015 article in *Psychology Today*:

"We have an evolutionarily ancient and highly interconnected pleasure circuit in our brains," Linden wrote, adding "This evokes the feeling of pleasure from both our vices (eating food when hungry, having an orgasm, drinking alcohol) and our virtues (meditation, learning, giving to charity).

"Here are the key findings that help to explain the pleasure-pain connection. When subjects in a brain scanner received an injection into the jaw muscles that produced a protracted aching type of pain, this triggered dopamine release in the nucleus accumbens, and the greatest release was seen in those subjects who rated the pain as most unpleasant. In rats, one can examine this phenomenon in greater detail. Electrical recordings from single dopamine neurons of the ventral tegmental area revealed that all of these neurons responded to the presentation of a tasty sugar droplet, yet some of these neurons responded to a brief painful foot shock with a decrease in their ongoing rate of activity while others responded with an increase.

"In other words, these latter dopamine-using neurons were salience detectors, releasing dopamine in response to either pleasure or pain. We also know, from different experiments, that protracted physical pain and protracted emotional pain (resulting from social rejection) can cause the release of endorphins, the brain's own morphine-like molecules, and that these endorphins can activate dopamine neurons in the ventral tegmental area. The end result is that there is an innate rewarding component to both pleasurable and painful experiences.

"How then, can we account for individual differences? Why do surveys reveal that only 5 to 10 percent of people enjoy receiving pain in a sexual context? The short answer is that we don't entirely know. Understanding how sexual kinks develop has not been a funding priority for government agencies and biomedical research charities. There are variant forms of dopamine receptor genes that attenuate the experience of pleasure and increase risk-taking and novelty-seeking behavior. However, it's not clear that these gene variants or any others (such as those related to endorphin signaling or pain perception) are linked to the practice of sexual masochism."[22]

In a 2015 article for the journal *Grey Matters*, Kathryn Stangret and Kat Ramus delineated the differences between masochism and self-harm.

"Masochism in general is defined as engaging in activities that lead to otherwise avoidable suffering. The most well-known form of masochism is found in a sexual context and includes deriving sexual gratification from being physically or emotionally abused.

"Perhaps one of the reasons sexual masochists are differentiated from the general population is because they perceive pain differently. Two studies done by Defrin, Kamping, and their colleagues compared people who routinely practice sexual masochism and people who do not. Both studies used questionnaires to assess masochistic activities, context-related pain experiences and emotions, pain catastrophizing, and fear of pain. In addition, Defrin used pressure tools to assess participants' pressure pain thresholds while Kamping used functional magnetic resonance imaging (fMRI) and photos of people to study brain activation patterns. Sexual masochists were found to have higher pain thresholds than controls, and their pain tolerance increased with the number of sessions and number of body parts involved in their masochistic behavior. This increased pain tolerance is termed hypoalgesia. However, sexual masochists' attitudes toward pain were determined to be context-dependent. They report reduced pain intensity and unpleasantness when painful stimuli were presented in a sexual context and not in a neutral context."[23]

With self-harm, however: "Although people who self-harm use pain as an outlet and to regain a sense of control similar to the way sexual masochists do, they do not necessarily enjoy the pain they experience. The control over their own pain satisfies the desire for autonomy over their experiences. This is an attempt to avoid uncontrollable, greater suffering, by willingly subjecting oneself to an autonomous form of suffering. This need for control likely stems from a history of trauma or mental illness. Because self-harm often occurs with mental disorders like depression, people may feel a lack of autonomy over their mental state and emotions and may seek to regain some of that control by projecting their emotional pain into physical pain."[24]

Another theory holds that masochism and self-harm may not be choices as much as obedience to a carved pathway in the brain.

"A clinical observation suggests a hypothesis that differs from classical analytic Oedipal theory as to the origin and psychodynamics of a type of male sexual masochism. This type of masochism may occur when a painful childhood life experience regarding severely forbidden sexual pleasure is associated and amalgamated with shame, humiliation, and feared physical and psychological punishment resulting in sexual pleasure. If this takes place with sufficient frequency during the critical phases of childhood sexual development, possibly including early adolescent sexual development, it becomes recorded as long-term memory in brain neural networks.

"In later years, the man feels compelled to reproduce in a masochistic ritual the former childhood psychological and physical condition, to bring about the most intense sexual pleasure. Humiliation, shame, discomfort, helplessness, and even possible pain are simulated or instigated. By voluntarily placing himself in charge of the type of sadistic treatment he receives in acts of masochism, a patient may unconsciously place himself in charge of his plight, in an attempt to master it, in contrast to his childhood helpless state. This formulation does not require an Oedipal explanation."[25]

Interestingly, that would seem to contradict other brain memory pathways that warn against painful situations. That may be one reason why the article in *Psychology Today* clearly classifies masochism as a "disorder."

"Sexual masochism disorder falls within the category of psychiatric sexual disorders known as paraphilias, which involve recurrent, intense, sexually arousing fantasies, urges, or behaviors that are distressing or disabling and have the potential to cause harm to oneself or others. Sexual masochism refers to engaging in or frequently fantasizing about being beaten, bound, humiliated, or otherwise made to suffer, resulting in sexual satisfaction. If people with this sexual preference also report psychological or social problems as a result, they may be diagnosed with sexual masochism disorder. The types of distress that people with this disorder may experience include severe anxiety, guilt, shame, and obsessive thoughts about engaging in sexual masochism. (If a person has a masochistic sexual interest but experiences no distress and is able to meet other personal goals, then they would not be diagnosed as having a disorder.)

"One specific type of sexual masochism is called asphyxiophilia, in which a person receives sexual satisfaction by having their breathing restricted. While some people engage in this practice with partners, others prefer to restrict their breathing while they are alone, and accidental death may happen as a result."[26]

When a masochist becomes tied up (no pun intended) with someone with sadistic tendencies, the result can be hellish.

As I write this, a case is playing out in the courts involving a well-known Major League Baseball pitcher and the woman he is accused of injuring during rough sex. Evidence shows that on an internet site, the woman requested being hurt, only to be hospitalized when the person who responded inflicted various injuries on her. Nevertheless, she then invited him to participate in another session, only to be injured again, at which point she filed for a restraining order.

The *Psychology Today* article postulates that masochism and self-harm become a psychiatric disorder when these activities transcend choice or "quirkiness" and become either dangerous or sources of intense guilt.

"To be diagnosed with sexual masochism disorder, according to the DSM-5, a person must experience recurrent and intense sexual arousal from being beaten, humiliated, bound, or from some other form of suffering. These types of urges, fantasies, or behaviors must be present for at least six months and cause clinically significant troubles or difficulty in social, occupational, or other important areas in life."[27]

Drawing a line between recreation and compulsion evokes similarities between masochism and drug abuse. In the latter case, it is entirely possible that escalating the amount and strength of drugs ingested might reflect an innate desire for physical harm.

"A multitude of interconnected factors was theorized to alter the experience of BDSM pain, including: neural networks, neurotransmitters, endogenous opioids and endocannabinoids, visual stimuli, environmental context, emotional state, volition and control, interpersonal connection, sexual arousal, and memories. The experience of pain in this context can bring about altered states of consciousness that may be similar to what occurs during mindfulness meditation. Through understanding the mechanisms by which pain may be experienced as pleasure, the role of pain in BDSM is demystified and, it is hoped, destigmatized."[28]

Another line of thinking is that sexual masochism is not an entity unto itself, but is only tied into what has been termed "moral masochism."

"Some individuals may unwittingly set themselves up for repeated failures in work and love relationships because of an unconscious (or implicit) need to punish oneself. For example, an individual may get themselves very close to a promotion at work but then blow it in a way that looked inevitable and seemingly planned.

"The implicit need to punish oneself satisfies pervasive, extreme, oppressive feelings of guilt. Guilt is the predominant feeling among what traditional psychoanalysis labeled 'moral masochism' (as opposed to sexual masochism the fetish). Another way of describing the pattern of unconsciously arranging to get close to success but ultimately not to make it is to say one is engaging in 'self-sabotage.'"[29]

"The self-sabotaging character has been labeled 'masochistic' in psychoanalytic theory. Prior to describing the psychodynamic conceptualization of masochism further, it is necessary to point out that psychodynamic theory posits the existence of unconscious processing which can include motivation and emotions that we are unaware of."[30]

If nothing else, there seems to be a consensus among both psychiatric and medical researchers that masochism is a uniquely human trait.

Animals have been trained to self-harm, but only by "positive reinforcement," in which animals are taught to associate pain with a reward. "Generally, when an animal experiences something negative, it avoids it," explains Paul Rozin from the University of Pennsylvania.

"Benign masochism is something that those who engage in BDSM won't find surprising. Mistress Alexandra, a professional sadist based in London, explains: 'We make a difference between good pain and bad pain. Bad pain indicates that something is not right, something we have to pay instant attention to. Then there's good pain which is enjoyable. For example, when the shoulder starts pulling during bondage, that's potentially unsafe so we release it.'

"The theory is also thought to explain why we seek out and enjoy other intrinsically unpleasant experiences, such as fear-inducing rollercoasters or sad movies. 'If an animal took a rollercoaster ride it would be scared, and it would never go again,' says Rozin."[31]

Or perhaps if humans were less constricted and guilt-ridden about sex, there would be far fewer masochists. Writes psychology researcher Orli Dahan: "I argue that sexual masochism and its related behaviors are healthy (and not a psychopathology) out of an evolutionary point of view, which is consistent with positive sexuality. Historically, the psychological literature concerning sexual masochism was negatively oriented—thus, many people actually felt (or still may feel) guilt and shame regarding their sexual masochistic desires and behaviors."[32]

Just another mystery offered up by the endlessly fascinating human mind.

Chapter 9

Keeping Pain in Mind

Mudhasir Bashir, MD

The recognition of chronic pain as a multifaceted condition appears to be bringing the healing community together—or at least onto the same page.

Medical care in the United States had begun to resemble an archipelago of small islands, each occupied by an individual specialist. Some overlap occurred in hospital settings, where different modes of treatment were available in one place, but general practitioners facing symptoms beyond their experience spent much of their time deciding which island to recommend.

Management of chronic pain now considers therapy from pain psychologists as important as medication management. Not so long ago, if a doctor told a patient, "I think you need to see a psychiatrist," it was considered disparaging. Now, it's often the admission of a complicated process that is better managed with a deeper understanding of the issues that might be leading to resistant pain.

The expression of pain is connected not to just the organ involved but also to the pathways and central control in the brain. All of our previously held medical beliefs don't always apply.

Inside that three-pound mass of neurons lie the secrets to a condition that affects a majority of people at some point in their lives. Moreover, almost everything we now discover about the brain and pain leads us to something else, sounding a call for diverse healing disciplines to join forces.

"Pain is one of the most difficult medical problems to diagnose and treat and can be a common symptom of several psychiatric disorders. Pain-related issues are heterogeneous and often underestimated or misinterpreted, with the result that psychiatric interventions, which might have been beneficial from the outset, are often delayed or requested only as a last measure."[1]

Nevertheless, the wall between "imaginary" and "real" pain is gradually crumbling. For if pain can be imagined—or, in some cases, remembered—and it still hurts, then perhaps it can also be reimagined or forgotten.

Such thinking is the stuff of science fiction. It is difficult enough to subscribe to the theory that pain happens only after alerts dispatched by nociceptor neurons reach some automatic department elsewhere in the brain and quite another to believe that the recipient of those nociceptor messages actually considers a few alternative choices—without our conscious knowledge—before acting.

We marvel at the incredible speed at which a GPS system can spit out the exact distance between an address in downtown Miami and an intersection in rural North Dakota, or our computerized Scrabble game can hand down its irrefutable decision on a word choice.

The brain is even faster.

We have also learned that thoughts and memories can create "tracks" in the brain that can later come back to haunt us—or help us. Everything we have ever said or done is filed away in there, making the brain a repository of information about each individual life. True, not all of that information can be retrieved on command, but it's there.

When dealing with chronic pain, physicians still encounter cases that seem to defy a medical solution. This is where treatment can grind to a halt, reduced to a never-ending stream of drugs. It is also where the explorers of the mind can step in.

Medicine deals with pain from the outside in, psychiatrists and counselors from the inside out. If the physical symptoms seem to be linked, there may still be room for a change in mindset and attitude.

Rene Descartes opened the door to a new way of seeing ourselves when he declared, "I think, therefore I am." Now, the operative mantra in chronic pain treatment and research might be "I think, therefore I hurt."

Yet since pain cannot be objectively measured and is a very subjective thing, it is impossible to truly quantify, and it affects both the body and the mind. Medical and mental health professionals previously would tend to approach it from opposing directions.

Unable to get at the origins of a chronic pain problem, physicians often settled for simply trying to lessen the hurt. In contrast to this, psychiatrists and counselors often saw the outward symptoms as secondary and wanted to identify what was causing them. Either way, patients were only partially served.

Certainly, the current opioid epidemic has dispensed with the idea that individuals with apparently untreatable chronic pain could hold it at bay forever with the right medications.

In a 2014 article for the publication *General Hospital Psychiatry*, Catherine Q. Howe and Mark D. Sullivan wrote: "Chronic pain is highly comorbid with common psychiatric disorders. Patients with mental health and substance abuse disorders are more likely to receive long-term opioid therapy for chronic pain and more likely to have adverse outcomes from this therapy. Although opioids may exert brief antidepressant and anxiolytic effects in some patients with depression or anxiety, there is scant evidence for long-term benefit from opioid treatment of psychiatric disorders." Psychiatry, they added, is "the missing P in chronic pain care."[2]

Thus, a chicken-or-egg question has been raised: does chronic pain create mental health issues, or do mental health issues lead to chronic pain? The answer, however frustrating, is all too often "It depends."

According to *Current Psychology Reports*: "Treatment of chronic pain and comorbid mental health issues requires a multidisciplinary approach. Advancements in how pain is understood, especially centralized pain, have helped inform both pharmacological and behavioral interventions for pain. Given the growing concerns about the opioid epidemic and the lack of data supporting the use of opioids for long-term pain management, new treatment approaches are needed.

"We believe psychiatrists are an important piece of the pain management puzzle."[3]

When medical journals first began speculating about what would come to be known as chronic pain, many physicians were initially dubious that it was truly a separate physical condition. Since the diagnostic tools they had come to believe in told them that there was nothing identifiably wrong with many of their chronic pain patients, they often attributed the problem to depression, hypochondria, or some other psychiatric or emotional factor.

This, as it turned out, missed the point. Whatever its origins, the pain was real.

Meanwhile, psychiatric instruction continued to address the mental and emotional issues inherent in pain management rather than the pain itself. This omission, said Boston-based psychiatrist Igor Elman, was counterproductive.

"Pain problems are exceedingly prevalent among psychiatric patients," Elman wrote. "Moreover, clinical impressions and neurobiological research suggest that physical and psychological aspects of pain are closely related entities. Nonetheless, remarkably few pain-related themes are currently included in psychiatric residency training."[4]

Even when psychiatry and medicine draw nearer to an agreement on chronic pain, they still rarely collaborate on individual cases.

"The problem is that these patients fall between the cracks," said Elman. "Psychiatrists don't want to deal with these issues because they don't know much about pain, whereas our [other] medical colleagues don't like treating psychiatric patients because there is stigma attached to them."[5]

Part of this physiological-psychiatric rift may have stemmed from one of the few occasions in which mental health treatment edged into physical intervention. "Lobotomy" was an umbrella term for a series of different operations that purposely damaged brain tissue in order to treat mental illness, said Dr. Barron Lerner, a medical historian and professor at New York University Langone Medical Center in New York.

Doctors first began manipulating the brain to calm patients in the late 1880s, when the Swiss physician Gottlieb Burkhardt removed parts of the cortex of the brains of patients with auditory hallucinations and other symptoms of schizophrenia, noting that it made them calm (although one patient died and another committed suicide after the procedure), according to Encyclopedia Britannica.[6]

"The Portuguese neurologist António Egas Moniz is credited with inventing the lobotomy in 1935, for which he shared the Nobel Prize for Physiology or Medicine in 1949 (later, a movement was started to revoke the prize, unsuccessfully)."[7]

"'The behaviors [doctors] were trying to fix, they thought, were set down in neurological connections,' Lerner said. 'The idea was, if you could damage those connections, you could stop the bad behaviors.'"[8]

Procedures for lobotomies varied, including some methods that bordered on the bizarre. "Italian psychiatrist Amarro Fiamberti first developed a procedure that involved accessing the frontal lobes through the eye sockets, which would inspire

American neurosurgeon Walter Freeman to develop the transorbital lobotomy in 1945, a method that would not require a traditional surgeon and operating room. The technique involved using an instrument called an orbitoclast, a modified ice pick, which the physician would insert through the patient's eye socket using a hammer. They would then move the instrument side-to-side to separate the frontal lobes from the thalamus, the part of the brain that receives and relays sensory input."

The chaotic state of many mental institutions was at the center of this trend. By giving unruly patients lobotomies, doctors could maintain control over the institution, Lerner said.

"That's exactly what happens in the 1962 novel and 1975 film *One Flew Over the Cuckoo's Nest*, in which Randall Patrick McMurphy, a rambunctious but sane man living in a mental hospital, is given a lobotomy that leaves him mute and vacant-minded."[9]

People with mental illness were often vulnerable to such experimental tactics, having been largely abandoned by their families in many cases. Too often, the decision of whether to perform a lobotomy rested entirely with the hospital staff.

Weighed down by a host of negative reports, lobotomies eventually fell out of favor. Meanwhile, many psychiatric hospitals and private practitioners began to employ electroconvulsive therapy (ECT), commonly known as shock therapy.

According to a position paper from the American Psychiatric Association: "Extensive research has found ECT to be highly effective for the relief of major depression. Clinical evidence indicates that for individuals with uncomplicated, but severe major depression, ECT will produce substantial improvement in approximately 80 percent of patients. It is also used for other severe mental illnesses, such as bipolar disorder and schizophrenia. ECT is sometimes used in treating individuals with catatonia, a condition in which a person can become increasingly agitated and unresponsive."

On the downside, however, "like any medical procedure, ECT has some risks. ECT treatment has been associated with short-term memory loss and difficulty learning. Some people have trouble remembering events that occurred in the weeks before the treatment or earlier. In most cases, memory problems improve within a couple of months. Some patients may experience longer-lasting problems, including permanent gaps in memory."[10]

Lobotomies and ECT were not specifically associated with the treatment of either acute or chronic pain, but their negative aspects caused many critics, both in the medical profession and the general public, to conclude that psychiatry had no business assuming a medical role. This contributed to the duality of pain treatment, a split that has traditionally been physical as well as philosophical.

In most hospitals, the medical and psychiatric sections remain separate, and there appear to be few meetings on patient care attended jointly by medical and mental health professionals. It is also rare to have private practice physicians and psychiatrists sharing the same office space.

This anomaly was addressed in 2010 in a *Psychiatric Times* article, along with some suggestions:

"There are two concepts implied by the term 'integration.' The first concept is structural integration, which refers to the way physical space has been allocated within primary care offices; this facilitates collaboration, purposefully or not, between mental health and primary care. This can vary considerably, with some psychiatry offices located within primary care space (complete structural integration) and others located on a separate floor or in a separate building from primary care (no structural integration).

"The second concept is coordination, which refers to standardized approaches that are used to assign activities and responsibilities in advance of the performance of services, to specify the types of services being provided and the skills required to perform the services, and to facilitate the transfer of information from one person to another. Coordination can be increased by changing management structures. However, these strategies are often challenging to implement, especially in the private sector."[11]

Another possibility, greatly accelerated by the COVID-19 pandemic, is to make creative use of internet options such as Skype and Zoom to interact with patients.

"Experts had predicted for years that the field's most intimate treatment—psychotherapy, or the talking cure—was poised to go largely virtual, for many or most patients, forever altering day-to-day practice. In this extraordinary year, they are likely to be proved right."[12]

In March 2021, federal health officials loosened restrictions on practicing across state lines and began to expand reimbursement. Clinics across the country went virtual, with most consultations done by phone or computer. The number of virtual mental health visits in the sprawling Veterans Affairs health systems jumped more than sevenfold, from 7,500 to 52,600, in just the first two months of the US epidemic.

"Overnight, everyone began practicing telepsychiatry," said Dr. John Torous, director of the digital psychiatry division at Beth Israel Deaconess Medical Center, a Harvard affiliate."[13]

Even post-pandemic, this could be adapted to the issues around chronic pain in several ways. For both physicians and psychiatrists/counselors, it provides a means to reach out to chronic pain patients on the days when they might be prevented by their condition from traveling to an office visit (or, in case of a crisis, on weekends). It could also make interaction easier with individuals who are reluctant to be around other people.

In the process, health practitioners could streamline many of the "just checking to see how you are doing" follow-up appointments that don't necessarily require face-to-face contact but take up valuable time for both patient and doctor when carried out in person.

"If widely used approaches such as cognitive-behavior therapy lose something crucial by being virtual, it is not evident from the studies done so far. In one study, for instance, a team led by researchers based at the Baltimore Veterans Affairs Medical Center tracked more than 100 veterans being treated for depression over six months, half of them engaging in traditional, in-person therapy, the other half receiving care online. Both groups improved, on standard measures, by the same amount.

"Another study, led by Leslie Morland of the Department of Veterans Affairs, Pacific Islands Healthcare System, compared in-person and virtual talk therapy for 120 veterans with post-traumatic stress. It reached a similar finding: improvement across the board, no difference between the groups."[14]

These recent developments are relevant to this chapter because they point to possible new areas of cooperation between psychiatrists and physicians. Given the suggestions made in *Psychiatric Times*, it isn't unrealistic to imagine representatives from these two formerly diverse entities working together on a single case and even interviewing the patient together on Zoom (or sharing a Zoom recording after the fact).

The mutual benefits are obvious. For the psychiatrist, it is much easier to deal with a patient who is not distracted by intense pain. For the physician, it has already been demonstrated that a person's mental outlook can affect the amount of pain he or she feels.

One major obstacle in the path of psychological and medical cooperation has been, not surprisingly, the reluctance of insurance companies to recognize it.

"Unfortunately, the U.S. healthcare delivery system is rife with inconsistencies that may inadvertently push the gradient to the prescribing of pharmaceuticals in the absence of sufficient reimbursement for interdisciplinary care. In many countries around the world, the availability of interdisciplinary programs is increasing. In contrast to other industrialized nations, according to an analysis by Michael Schatman, Ph.D., published by the International Association for the Study of Pain, the United States now has only one interdisciplinary program for every 670,000 patients with chronic pain.

"Further difficulties with current coverage include the limited time allowed to provide comprehensive services and the inconsistency of available benefits to all patients with chronic pain. Lack of uniformity, for example, was shown in the final rule for the Final Fee Schedules for Physicians and Ambulatory Surgical Centers issued by the Centers for Medicare and Medicaid Services (CMS), which featured significant cuts to physician payment when they perform epidural injections in the office setting and a 20% increase in payments for the same services performed in hospital outpatient departments."[15]

Some entities, such as Medicare and Medicaid, and those policies included under the Affordable Care Act, have at least made an effort to include coverage for chronic pain. Yet Will Rowe, CEO of the American Pain Foundation, has seen the other side all too often.

"In the last few years, we've had a dramatic increase in people calling up and saying, 'I need help, my insurance company won't cover my pain treatment,' Rowe said."[16]

Certain issues are particularly problematic.

"'Insurance will cover only certain parts of a patient's pain management therapy,' Rowe said. 'They might not cover the physical therapy, but they'll cover the medicine. Or, if they cover the physical therapy, they only cover three sessions.'"[17]

Insurance also forces people to go through unnecessary steps before approving the recommended therapies that will manage pain for them, Rowe added, giving as an example the difficulty involved with getting certain FDA-approved medications for fibromyalgia.

"In many instances, insurers are saying you can't get those medicines until you go through and fail with four or five other treatment options. It might be that medicine is exactly what you need, but you have to go through four or five months of trial and error before you end up getting what you needed in the first place."[18]

Many insurance companies always have been resistant to chronic pain claims, which coincide with problems experienced by individuals seeking coverage for mental health treatment in general.

In one sense, it's all about the diagnosis. If someone receives treatment for a broken arm or a kidney stone, the cause is obvious. In many cases related to mental health, however, arriving at a short-term diagnosis can be difficult.

This is problematic on several levels for those with chronic pain.

In many instances, their injury or illness is either keeping them out of work or has forced them to quit altogether. With that income stream eliminated or trimmed down to unemployment benefits, it becomes difficult or impossible to pay for psychiatric care without the guarantee of insurance reimbursement.

Furthermore, the longer it takes for treatment to begin, the greater the possibility that a patient will become more depressed or anxious or develop a dependence upon the opioid medications that insurance companies often *will* pay for.

Many insurers, realizing that addiction treatment in a rehabilitation center or hospital setting is an expensive proposition, have pared down what they were willing to pay for opioid drugs. At the same time, though, they still consider many alternative methods of pain treatment to be "experimental," no matter how long they have been in use. These two seemingly contradictory strategies can squeeze chronic pain patients in a catch-22 vise.

"'The epidemic isn't just about how easy opioids have been to come by. It's also about how hard it is to access alternatives,' said Caleb Alexander, co-director of Johns Hopkins University's Center for Drug Safety and Effectiveness. 'No one ever died of an overdose of physical therapy.'"[19]

"We need to move a mountain," added Kate Berry, senior vice president for clinical affairs and strategic partnerships at America's Health Insurance Plans, the leading industry group. Berry and others also pointed out that pain is subjective and that providers often have differing opinions about how to effectively treat patients.

"You could ask 10 different doctors or physical therapists, 'How many visits should it take for somebody to resolve an intermittent pain in their neck?' You're going to get different answers," Berry said.[20]

Patients may also find it challenging to figure out what pain treatments insurers will pay for and what limits might be attached to that coverage. A 2016 study by the American Occupational Therapy Association found roughly half of all Obamacare plans failed to clearly indicate how many physical or occupational therapy sessions insurers would pay for before coverage ran out.

"Transparency is low," said Rebecca Peters, a research associate at the Urban Institute's Health Policy Center. "It's opaque and difficult even if things are covered."[21]

In 2015, the American Medical Association passed Resolution 112-A-14, submitted by the American Academy of Pain Medicine, which asked that "our American Medical Association (AMA) and interested stakeholders advocate for a minimum set of health insurance benefits for people in pain severe enough to require ongoing therapy. At a minimum, a proposed program of treatment categories should include:

(1) Medical management; (2) Evidence or consensus-based interventional/procedural therapies; (3) Ongoing behavioral/psychological/psychiatric therapies; (4) Interdisciplinary care (Evidence-based complementary and integrative medicine [e.g., yoga, massage therapy, acupuncture, manipulation]).[22]

What mental health issues go along with pain, especially chronic pain? We can start with the largest and fiercest elephant in the room—suicide.

In 2018, according to one study, "Chronic pain was present in 9% of 123,000 people who died by suicide. The rate rose over time from 7% to 10%, while suicide rates from opioid overdose remained the same (2%). The most common medical conditions were back pain, cancer, and arthritis. Numerically, suicide decedents with pain were older, more likely to be female than those without pain and to have depression (82%; 75% in decedents without pain) or anxiety (19% and 10%), and less likely to have intimate partner problems (17% and 32%). Firearms were the most common suicide method in over 50% of both groups. However, a greater percentage of chronic pain decedents died of opioid overdose (16% and 4%). Of 200 decedents with suicide notes that were reviewed, over half indicated that pain played a role in their decisions to die."[23]

This is hardly surprising. Many of those with chronic pain are alone, especially the elderly, and enforced isolation tends to curdle the mental outlook and erode resilience.

In an article published in the journal *Pain Medicine*, behavioral scientist Martin Cheatle speculated that many of the deaths attributed to accidental overdoses may actually be suicides.

"Comorbid conditions that pose risks for suicide, especially depression, are prevalent in people living with chronic pain. The true numbers of failed attempts and successful suicides are unknown and may never be determined. Yet risk factors for suicidal ideation are so high in this population that it must be assumed that some proportion of those who die of drug overdoses might have intended to end their lives, not just temporarily relieve their pain."[24]

Even those who are not alone may find themselves increasingly at odds with those around them. Sleep is frequently disrupted, adding to the emotional burden of fatigue. And because chronic pain is, by its very nature, stretched out like a rubber band over time, visits from friends and relatives often taper off.

"Being in constant pain is a horrible experience. Knowing that this pain is long-term, and for some people constant, can make it feel impossible to go on. It's completely natural not to want to live in pain. Your body and mind are not designed to cope with being in pain regularly. When patients are not given another way to escape from their pain (meaning through treatment), they can feel that there is only one escape route."[25]

One of the examples in Cheatle's article was a woman referred to as Mrs. D, "a 45-year-old mother of 2 who was actively involved in the care of her family, volunteered in the community and worked full time as a para-educator at the local high school. She had no history of mood disorder and had never sought mental-health treatment when she began to note left big toe pain. An evaluation by a podiatrist led to a tarsal tunnel release, which provided no relief. Instead, her pain spread to both feet. She entered pain management and underwent a series of bilateral L4 lumbar sympathetic blocks, which also failed to reduce her pain. She was subsequently seen by a host of medical specialists, locally and at tertiary facilities. Numerous nerve conduction studies revealed progressive peripheral neuropathies of an unspecified etiology. Further testing included a painful nerve biopsy, which yielded equivocal results.

"Meanwhile, her symptoms progressed to include both hands, and though formerly an avid reader, she was unable to even turn pages in a book. She ceased working and relinquished her volunteer positions. Her spouse assumed all her household duties, causing her to feel inadequate and guilty. Her treatment included a polypharmacy of opioids, antiepileptic drugs, antidepressants, hypnotics to induce sleep, and, eventually, a 4-month stay in a residential pain program. Despite these efforts, her pain worsened, her function deteriorated, and her mood plummeted. She began experiencing panic episodes and deepening depression, confiding to one of her treating physicians her fears of becoming wheelchair bound. She admitted to daily suicidal ideation and hoarding her prescribed medications, stating, 'If I can't take it anymore, I will kill myself.'"[26]

Melissa Wardlaw, who describes herself as a "chronic illness and pain peer support counselor," tells a similar story.

"There was a period of time in my medical journey for years—years ago—that I didn't want to be here anymore, for no other reason than I wanted the life-changing, soul-crushing pain to stop—at any and all costs, including death. I was 85 to 90 percent homebound/bedbound, couldn't drive and many times had to get used to using new assistive devices during the rare times I did get out (yet I looked and sounded relatively 'normal' considering I struggle[ed] with invisible chronic illnesses/pain). All this after a successful and lucrative career in business was stripped away from me in a split second by a routine medical procedure gone wrong. The grief and loss were palpable."[27]

In dealing with chronic pain, physicians must always consider a patient's mental state as well as physical condition. The problem is that there is a tendency in the medical profession to regard depression as merely an appendage to pain. Fix the pain, and the depression will improve automatically.

Sometimes, that's true. But depression can also become a major obstacle thrown in the path of treatment options, and there are times when the two afflictions can feed off the other, spiraling the patient downward.

An important part of the equation is whether the depression is situational or clinical. Chronic pain not only hurts physically but its impact ripples outward into all aspects of a person's life. Job performance is often affected, which may lead to unemployment. Activities that once relieved stress become difficult. Someone in serious chronic pain often becomes unable to contribute to housework or yardwork. Intimacy and sexual activity may cease.

In other words, pain that refuses to go away can expand to fill up a patient's life, crowding out everything that once gave it vibrancy and meaning. It's no wonder that a person becomes depressed—who wouldn't?

As Kathleen Crowley put it in a 1996 article for the National Pain Outreach Association, "The very nature of depression is at odds with the treatment of it. Severe depression can be defined in one word—hopelessness. A severely depressed person has no hope for the future and no memory of things ever having been any different. . . . Swallowing a pill may be the only action a severely depressed individual is initially capable of."[28]

According to a 2007 article in *Psychological Bulletin*: "Some have suggested that chronic pain is a form of 'masked depression,' whereby patients use pain to express their depressed mood because they feel it is more acceptable to complain of pain than to acknowledge that one is depressed."[29]

This has no doubt changed, to some degree, during the past twenty years. Depression has begun to creep out of the closet of hidden human conditions, especially since it has been publicly acknowledged by some high-profile celebrities.

For many, though, it remains a sign of weakness or at least something to be kept private. Even now, whenever we pass someone on the street and ask, "How's it going?" chances are they will reply, "Good!" or "Fine!" Probably not one in a hundred would respond with "Well, things aren't going so well for me right now. Can I tell you about it?"

In a June 2021 article for the *Guardian*, science correspondent Linda Geddes wrote: "Chronic pain is more common among people who experienced lots of surgery as children or emotional or physical abuse. There is also a strong link between chronic pain and depression. Of course, living with chronic pain can be depressing, but depression can also amplify pain processing in and of itself.

"This is not the same thing as suggesting that someone's mental attitude can affect their pain.

"'Rather, people often don't realize that the brain pathways that drive depression are intrinsically linked to the ones that drive chronic pain,' says Dr. Kirsty Bannister, a senior lecturer at King's College London, who researches these pain pathways.

"When pain signals arrive in the brain, they are transmitted to areas that allow us to locate and categorize that pain, as well as to areas that process emotions. These brain areas then send signals back down the spinal cord, which, in health, causes those original pain signals to be dampened. This is the reason your thumb begins to stop hurting several minutes after you've bashed it with a hammer.

"But in people with a history of depression or childhood trauma, those emotion processing areas are often wired differently, which can alter the type of signals they send. As a result, their pain pathways carry on being activated.

"'Because our life experiences are not equal, it means that previous issues such as fear, hopelessness or helplessness may escalate your pain to a very different intensity compared to mine,' Bannister says."[30]

The strong connection between chronic pain and depression has naturally led many physicians and mental health professionals to try antidepressant drugs on chronic pain patients. One report from the Mayo Clinic sounded promising:

"Some of the more effective and commonly used medications for chronic pain are drugs that were developed to treat other conditions. Although not specifically intended to treat chronic pain, antidepressants are a mainstay in the treatment of many chronic pain conditions, even when depression isn't recognized as a factor."[31]

The Mayo report noted that antidepressants seem to work best for pain caused by:

- Arthritis
- Nerve damage from diabetes (diabetic neuropathy)

- Nerve damage from shingles (postherpetic neuralgia)
- Nerve pain from other causes (peripheral neuropathy, spinal cord injury, stroke, radiculopathy)
- Tension headache
- Migraine
- Facial pain
- Fibromyalgia
- Low back pain
- Pelvic pain
- Pain due to multiple sclerosis

"The pain-killing mechanism of these drugs still isn't fully understood. Antidepressants may increase neurotransmitters in the spinal cord that reduce pain signals. But they don't work immediately."[32]

Certainly, depression often becomes an obstacle to dealing with (and treating) chronic pain. Every physician and psychiatrist wants his or her patients to be motivated partners in any plan for their healing and recovery, but a severely depressed individual might not be capable of participating to that extent.

"Major depressive disorder (MDD) is a multifaceted disease that presents with both emotional symptoms (e.g., depression, guilt, suicidal ideation) and physical symptoms (e.g., sleep disruption, gastrointestinal disturbance, unexplained aches and pains). Headache, neck and back pain, abdominal pain, and musculoskeletal pain are common in patients with depression."[33] When a patient presents with depression and chronic pain, it is important to address both conditions carefully and thoroughly.

"The convergence of depression and pain is reflected in the circuitry of the nervous system. In the experience of pain, communication between the body and brain goes both ways. Normally, the brain diverts signals of physical discomfort so that we can concentrate on the external world. When this shutoff mechanism is impaired, physical sensations, including pain, are more likely to become the center of attention. Brain pathways that handle the reception of pain signals, including the seat of emotions in the limbic region, use some of the same neurotransmitters involved in the regulation of mood, especially serotonin and norepinephrine. When regulation fails, pain is intensified along with sadness, hopelessness, and anxiety. And chronic pain, like chronic depression, can alter the functioning of the nervous system and perpetuate itself."[34]

Depression is not the only mental health issue associated with chronic pain. Anxiety is equally common.

"In the case of chronic pain, the anxiety and avoidance behaviors become chronic themselves. The chronic anxiety leads to a chronic sense of alarm or distress, which makes patients edgy. Cognitively, it leads to a chronic focus on pain, which preoccupies the attention of the pain sufferer. Everyday decisions seem to turn on how much pain the patient has at any given time. It also leads to chronic muscle tension, which in turn leads to more pain. Chronic avoidance behaviors subsequently lead to an increasing sense of social isolation, inactivity, de-conditioning and, ultimately, disability.

"The common denominator between chronic pain and chronic anxiety is the nervous system. The nervous system has become stuck in a persistent state of reactivity."[35]

At its worst, chronic pain refuses to be ignored. Even if its symptoms have momentarily subsided, the uncertainty of when they might return is enough to keep anxiety at a high level. Many chronic pain patients talk of waking up every morning to find the same questions waiting for them: Will I be in pain again today? How bad will it be? Is it getting worse? Do I have enough medicine to get me through it?

The cruel irony around health-related anxiety is that it often makes its victims even sicker. Anxiety equals stress, which can equal stomach cramps, migraine headaches, and a whole phalanx of other physical ills over and above the chronic pain that triggered them in the first place. In that sense, anxiety becomes an accomplice to chronic pain, a partner in what seems a conspiracy to rob that person of any sense of control.

For psychiatrists, psychologists, and counselors, helping a chronic pain patient regain that control is often a necessary first step in treatment. From the mental health perspective, the nature of the pain is less important than its mental and emotional side effects.

"One of the most effective psychological approaches for pain management is based on the cognitive-behavioral therapy (CBT) approach. CBT is an effective psychological treatment for chronic pain that is aimed at changing maladaptive thoughts and behaviors that serve to maintain and exacerbate the experience of pain.

"Key components of CBT for chronic pain include cognitive restructuring (e.g., teaching patients how to change maladaptive thoughts); relaxation training (e.g., diaphragmatic breathing, imagery); time-based activity pacing (i.e., teaching patients how to be more active without overdoing it); and graded homework assignments designed to decrease patients' avoidance of activity and to reintroduce a healthy, more active lifestyle. CBT also focuses on promoting patients' increased activity and functioning using techniques such as exercise homework, activity scheduling, and graded task assignments. A substantial literature exists documenting the efficacy of CBT for chronic pain conditions."[36]

Discussions about chronic pain and anxiety also often turn to the use of medical marijuana.

"Chronic pain relief is by far the most common condition cited by patients using cannabis for medical purposes. Studies and observational reports have demonstrated that more than 80% of patients enrolled in clinical trials and/or with a medical cannabis card indicate 'severe pain' as the reason for use. In addition, evidence proposes that some individuals with chronic pain are substituting the use of opiates with cannabis. Data from a survey directed by a Michigan medical cannabis dispensary claimed that use of medical cannabis was associated with a 64% decrease in use of opioids. In addition, a recent analysis of prescription data from Medicare Part D enrollees in states with medical access to cannabis suggested a substantial reduction in prescriptions for conventional pain medications, principally opioids."[37]

Not too long ago, marijuana was still identified primarily as a recreational drug. This presented some generational issues, especially among older people who had grown up thinking of it as illegal and dangerous.

"A growing body of clinical research and a history of anecdotal evidence support the use of cannabis for the relief of some types of chronic pain, including neuropathic pain, and spasticity associated with multiple sclerosis. In a recent comprehensive review of current data on the health effects of cannabis and cannabinoids, the National Academies of Sciences determined that adult patients with chronic pain who were treated with cannabis/cannabinoids were more likely to experience a clinically noteworthy reduction in pain symptoms. They rated these effects as 'modest.' Studies also imply some efficacy for cancer-related pain, migraines, fibromyalgia, and other pain conditions. It has been suggested that these compounds may be valuable in other conditions, including rheumatoid arthritis, osteoarthritis, and various other types of acute and chronic musculoskeletal pain. However, how different cannabis species, routes of administration, and doses differ in their effect is less clear, and more research is required.[38]

Another mental health/chronic pain connection worth noting concerns schizophrenia and congenital insensitivity to pain (CIP).

"Numerous reports indicate that, relative to normal, individuals with schizophrenia are insensitive to physical pain associated with illness and injury. In addition, insensitivity to pain of various sorts administered in experimental studies has been reported frequently in this population. This extensive and diverse literature of clinical and experimental reports suggests that many individuals with schizophrenia are less sensitive to pain than normal individuals. However, because the experimental studies—almost all of which were conducted before 1980—suffer from a variety of methodological limitations, this research provides neither a satisfactory characterization nor an adequate explanation of pain insensitivity in schizophrenia."[39]

Further examination finds strong anecdotal evidence that individuals with schizophrenia are less sensitive to pain than most of us, but this is not borne out by the majority of studies. Those in the "pro" camp have attributed this phenomenon to heredity, nociceptors that are somehow compromised by schizophrenia, and side effects from opioid drugs. Still, anything CIP-related interests researchers on the chance that it might lead to ways of turning off pain signals in general.

Whatever its manifestations and connections may be, however, chronic pain has moved far up the list of both medical and psychiatric priorities. The sheer number of people affected by it would make that a given, but there is also the potential for research that would not only change lives but reconfigure how we look at the relationship between body and mind.

Writing for *Psychiatric Times*, John D. Otis and Douglas H. Hughes suggested: "One way that psychiatrists can facilitate care related to pain is to organize a group of providers with specific expertise in pain management and rehabilitation into a multidisciplinary pain management team. Multidisciplinary pain programs consist of a group of providers with specific expertise in pain management and rehabilitation. The goal of the team is to provide a comprehensive pain assessment and deliver an integrated and coordinated plan for treatment that emphasizes medical, psychological, and rehabilitation approaches as needed. This type of approach is appropriate for patients with chronic pain that has been unresponsive to less intensive interventions."[40]

"Psychiatrists with an interest in working with persons who have chronic pain conditions should consider ways of integrating and coordinating services within the primary care setting, delivering services in an effective manner, and providing services as part of a pain management team. Psychiatrists with expertise in conducting comprehensive pain assessments and delivering evidence-based interventions for pain have an opportunity to make a unique and significant contribution to improving the quality of life for patients with pain."[41]

Many psychiatrists with a growing interest in the field of chronic pain are studying and using a diverse range of therapies to target chronic pain. These include older therapies like mindful meditation, CBT, and acceptance and commitment therapy. The newer therapies still in their infancy include pain reprocessing therapy (PRT) and emotional awareness and expression therapy (EAET). The newer ones follow a neuroscience-based psychological assessment approach and focus on the history of stress, emotional behavior, and brain changes caused by these. Though these have shown promising results, the ultimate success that translates into improving the quality of life for a patient can be achieved only when these therapies are routinely integrated into a multidisciplinary service at the primary care level that is easily accessible to all patients suffering from chronic pain.

Chapter 10

The Road Ahead

The longer humanity evolves and the more volatile the world we are creating becomes, the harder it will be to make any accurate long-term predictions.

Or even short-term predictions.

In 2017, a person like Donald Trump was one of the last people anyone thought would wind up being president of the United States. In 2019, who could have foreseen the rapid worldwide spread of a novel virus that would kill hundreds of thousands?

The 1960s "futurists" trying to envision the 1980s were mostly wrong because they didn't envision the rise of the internet. A few years later, the CEO of one of the nation's leading typewriter manufacturers tried to comfort his stockholders by saying: "Computers are just a fad. People will always need our typewriters." And Thomas Watson, a visionary and president of IBM, said in 1943, "I think there is a world market for maybe five computers."

After all, we can only make predictions based on now and the obvious signs of where that "now" may be headed. Yet there is always a wild card hiding in the tarot deck—the internet, COVID-19, Donald Trump—that shifts all those expectations.

If Rene Descartes could be transported via time travel from the seventeenth century to the twenty-first and shown how modern humanity has learned to confront its pain, he would probably say something like, "My God! This is hard to believe."

In seventeenth-century French, of course.

Certainly, more than a thousand years of asking questions and performing experiments have brought us a long way on that front. But talk to a current sufferer of fibromyalgia, shingles, or migraine headaches on one of their bad days, and they might assess those gains differently. For them, pain is still a regular visitor.

Before we figure out what is coming next, it makes sense to try to find out where we came from.

In the early 1990s, it looked like things were changing. Pain was designated as a "fifth vital sign" with the help of some rather suspicious elements in the medical and regulatory fields. Simultaneously, opioid medications like OxyContin were promoted as "safe" and "nonaddictive" treatments that would cure the world of pain. OxyContin was touted as resistant to abuse. Added to it was a high-end and aggressive promotion—based on a single letter to an editor—that claimed that less than 1 percent of people on opioid medication get addicted.

Providers who failed to treat pain "adequately" were penalized.

On the face of things, it falsely appeared that all of this was done to bring relief to millions in pain. However, it bought them misery and death.

We now know that it was all a facade. Carefully constructed code words including "individualize the dose," "double the dose safely," and, worst of all, the concept of "pseudoaddiction." This last designation falsely referred to addicts who sought more medications as being undertreated for pain so that they could, ironically, be prescribed more medications. Representatives from the medical community were being actively recruited for a fee by the drug companies to help them promote and sell more.

Many days each week, doctors were visited by multiple drug representatives extolling a so-called magic drug and lavished with expensive gifts, high-end dinners, and skiing trips.

We now have evidence that it was all meant to turn the pharmaceutical companies, mainly Purdue and its owners, the Sackler family of New York, into billionaires by ruthlessly pushing health-care providers to ignore any concerns and prescribe more and more. For creating all this tragedy, last year Purdue got a light slap on the wrist. It agreed to plead guilty to criminal charges and face penalties of roughly $8.3 billion for its role in the opioid epidemic.

A recent drama miniseries called *Dopesick* reveals the true story. The subheading after the title states: "They sold us a lie. We swallowed it whole." That is a largely accurate description of what happened. I know because I lived it, first as a trainee in anesthesia and pain management during the 1990s and then as a physician managing chronic pain and its complications for almost twenty years. I have followed the tragic trajectory since the beginning.

The result of all this was that even those who had been hesitant initially to prescribe opioids for non-cancer pain started prescribing and kept escalating the dosages. This was done based mostly on patient verbal feedback and not on any objective signs of pain. Many patients received medications in amounts that they would never tolerate under normal circumstances. Much of it was misused, abused, and diverted to the streets for profit and led to overdose deaths.

At one point in time, the United States consumed almost 80 percent of the oxycodone in the world, and almost 80 percent of that was dispensed or used in just one state: Florida and its so-called pill mills.

Obviously and deservedly, the pendulum has shifted. There is now more oversight of how much opioid medications are being prescribed by each provider. In turn, a more thorough screening and careful monitoring—including regular follow-ups, drug screens, and prescription monitoring—is utilized by the providers.

In spite of all this, the harm continues. Almost a hundred thousand people died of a drug overdose last year, many of which were related to prescription medications.

We now have highly advanced interventional pain management options available in the United States. Most chronic pain symptoms can be managed with a combination of physical therapy, regular physical activity, non-opioid medications, and interventional procedures. The latter include nerve blocks, muscle and joint injections, and peripheral, spinal cord, and brain modulation procedures. If employed properly, these can greatly reduce the need for opioid pain medications.

Still, pain researchers are waiting for that wild card. Although a list of effective medications now available for acute pain would take up several pages in this book, fending off chronic pain completely has proven to be much more challenging. Nevertheless, there is now hope where there was once only hurt.

One of the noteworthy aspects of medical research is that it has few secrets. It's hard to imagine a doctor or researcher coming up with a groundbreaking drug or technique then keeping it quiet. Such innovations are almost always reported in medical journals, presented at conferences, or revealed in some other public fashion, and that is certainly the case with chronic pain.

Each discovery, then, serves as a stepladder for those who follow. In some cases, individuals who have discerned part of a puzzle find fellow seekers examining another part. A detailed map of the brain has been created, and traffic along those microscopic byways is heavy.

We now know what we need to do, just not exactly how to do it.

True, the esoteric nature of chronic pain has attracted the medically curious from many disciplines. At the moment, explorations in this area are proceeding in a number of different directions:

1. The search to find new drugs that mimic the pain-fighting qualities of opioids like OxyContin but without the danger of addiction. Or, on a parallel track, ways of muting chronic pain without drugs.
2. New methods of manipulating the intricate communication among nociceptors, hormones, and the central brain.
3. Ways in which thought patterns can be adjusted to mute the effects of pain.
4. Using genetics to find out more about why certain people are afflicted with certain pain-related conditions.
5. Regenerative medicine, including new uses for stem cells.

All of these quests continue unabated, with some new development revealed nearly every week. At the same time, however, progress in actual treatment has also been affected recently by the COVID-19 pandemic and its impact on chronic pain and its management.

The arrival of the virus had two notable collateral effects on individuals dealing with chronic pain. The fear of being infected, especially before vaccines were unveiled, kept many of them from following through with their scheduled health-care appointments. And even if they decided to venture out and risk COVID-19, they often found that they had been bumped from the schedules of their providers by the flood of those who had been infected.

"Just like most other medical specialties, the field of chronic pain is one of the hardest hit by the COVID-19 pandemic, leaving many patients overburdened with their chronic pain with their ongoing treatment delayed. Over the past few months, in order to mitigate the spread of COVID-19, in-person access to physicians was limited and predominantly moved to telemedicine. While telemedicine has been beneficial in maintaining contact with patients and continuing therapy, due to an inability to perform a physical exam, it did delay and has created a lag in proper diagnosis and treatment. Furthermore, diagnostic imaging such as x-rays, CT scans, and MRI scans all have had to be deferred, which while certainly not required to make the diagnosis can be beneficial for diagnostic and therapeutic purposes.

"Additionally, due to deferment of elective cases, patients who were currently waiting to undergo extensive surgery such as laminectomy, discectomy, hip arthroplasty, or total knee arthroplasty have had their surgeries delayed and are having their pain managed with medicines, in some cases narcotics. Interventional pain procedures that needed multiple trials such as medial branch blocks, spinal cord stimulator trials have also been hindered further postponing the subsequent therapeutic procedures like radiofrequency ablations and spinal cord stimulator implants."[1]

Even worse, evidence has begun to accumulate suggesting that the virus might prove a contributor to the ranks of chronic pain patients rather than simply an impediment for them. For some individuals, so-called long-haul COVID, which continues for several months after hospital discharge, appears to bring with it symptoms strikingly similar to chronic pain.

"Although these infections have markedly disparate acute presentations, a stereotypical chronic syndrome occurred at remarkably similar rates and was not predicted by demographic, psychological/psychiatric measures or microbiological factors. The presence and severity of somatic symptoms during acute infection was closely correlated with the subsequent development of chronic fatigue and pain."[2]

In March 2021, it was reported that "a small team of French researchers suggests some patients with COVID-19 develop neuropathic pain within weeks or months following infection and that patients with neuropathic pain and COVID-19 sometimes present with deterioration of neurologic complications and/or pain exacerbation."[3]

And then, even more unsettling, there was this:

"Diagnostic imaging of some people who have had COVID-19 show changes in the brain's white matter that contains the long nerve fibers, or 'wires,' over which information flows from one brain region to another. These changes may be due to a lack of oxygen in the brain, the inflammatory immune system response to the virus, injury to blood vessels, or leaky blood vessels. This 'diffuse white matter disease' might contribute to cognitive difficulties in people with COVID-19. Diffuse white matter disease is not uncommon in individuals requiring intensive hospital care but it is not clear if it also occurs in those with mild to moderate severity of COVID-19 illness."[4]

A June 2021 article by Linda Geddes, a science correspondent for the *Guardian*, adopted a similarly pessimistic tone.

"The Covid-19 pandemic could make this situation even worse. One of the most common symptoms reported by people with 'long Covid' is musculoskeletal pain, and those with existing musculoskeletal pain conditions seem to be at risk of experiencing stronger pain after a serious Covid infection.

"'Musculoskeletal pain is an issue that we must start considering as a long Covid problem,' says Prof. Lars Arendt-Nielsen, immediate past president of the International Association for the Study of Pain (IASP) and director of the world's largest translational pain research center at Aalborg University in Denmark.

"The health profession as a whole does not treat people with chronic pain well. And it is about to be hit by a tsunami of severely ill people needing help."[5]

Arendt-Nielsen saw COVID-19 putting new demands on the nervous system of chronic pain patients, with potentially far-reaching effects.

"According to the old way of thinking, the body's pain pathways were arranged like a telephone switchboard, with fixed sets of wires (nerves) connecting our peripheral organs and tissues to the spinal cord, and further nerves connecting the spinal cord to the brain as well as feeding back down in the opposite direction.

"The assumption was that this was a fixed, solid, stable system," Arendt-Nielsen says. "But we now know that these neural networks can be reorganized when there are persistent inputs into the system, and cause an increased gain of the pain signal and hence generate a stronger pain."

As a professor of anesthesiology, psychiatry, and medicine at the University of Michigan, Daniel Clauw looks at chronic pain from a number of angles. What he sees now is a treatment system in flux as new information continues to surface.

"In many people with conditions like low back pain, endometriosis, irritable bowel, headache, fibromyalgia, you're not going to find a problem in the area of the body where the person is experiencing pain," he said. "There's more and more evidence that these are the central nervous system, systemic conditions, where the pain can present in different areas of the body at different points in time over that person's life."[6]

The *Guardian* article also described the case of fifty-five-year-old Vicky Naylor, a nurse from the United Kingdom, who "developed fibromyalgia after undergoing an emergency C-section 11 years ago, but through a combination of swimming, yoga, and medication, had her symptoms under control. 'It was only if I had an illness or got very stressed that I'd get flare-ups. I'd become very stiff, and get painful joints and trigger points [tender spots around the body] and headaches, but I could always get them back under control and deal with them,' she says. 'I never missed a day's work.'

"Then, in March 2020, she developed Covid-19. Her initial symptoms were severe—fever, shortness of breath, and a cough so bad she lost her voice—although she was never admitted to a hospital. But as the weeks turned to months, she began to experience a plethora of other symptoms, including excruciating pain. 'It is off the scale,' she says. 'I get it in my feet—some days it is so bad I can't bear to wear filled-in shoes. I get it in my knees, the tops of my legs, my right hip particularly, my elbows, shoulders, neck—it's just everywhere.'"[7]

By the time this book is published, more clarity might have emerged regarding some of these potential COVID–chronic pain connections. On the downside, forward movement in other directions will be delayed in some cases by the pressing need to address the more immediate links to COVID-19. One silver lining, however, is that merging these two research topics will attract the attention of a whole new group of medical investigators who had not previously focused on chronic pain.

Although the internet and current medical literature are bursting with news of promising developments in the pain field, "promising" is usually the operative term. Typical caveats include the need for further testing on humans rather than mice, uncertainty about possible side effects, and caution about the possible cost of deployment.

One suggestion does not come with hidden dangers or a high price tag, however. A number of treatment entities, including Veterans Affairs, have been making an effort to draw chronic pain patients more directly into their treatment plans. For those who acutely feel the loss of control, a return to decision making is a plus.

This has always been a blind spot for the medical community. In too many cases, individuals with chronic pain are offhandedly given an explanation of their condition that is almost impossible for them to understand then handed a prescription and sent home to ponder their fate—often in isolation and confusion.

Treatment for acute pain cases generally comes with a hopeful time line stretching out toward full recovery, complete with steps the patient can take to eventually achieve this goal. With chronic pain, it is more often some version of, "Well, try this and see if it works. If it doesn't, let me know and we'll try something else."

Psychologically, for many people, this is a dead end. All too often, chronic pain patients are not offered a cure, only abatement, leaving them with the uncomfortable feeling that their medication is the only thing keeping the wolf of pain from their door.

This is not to say that chronic pain conditions can't be cured or even disappear on their own at some point. But a discouraged patient is in danger of focusing nearly total attention on his or her affliction, which does nothing to facilitate healing.

One template that is being developed has the patient deciding on an individual plan for recovery—perhaps walking to the mailbox in the first week, going back to doing laundry the second, and so forth. People trapped in these circumstances desperately need to see progress.

"The current pandemic has exacerbated existing sources of social threat for people with chronic pain. To prevent a population-level increase in the severity and impact of chronic pain, it is critical to devote scientific attention to the assessment, mitigation, and prevention of sources of social threat for people with chronic pain. Crucially, COVID-19 should not only be regarded as a challenge but also as a unique opportunity for researchers and clinicians to develop new ways to deliver social support and pain management, as well as understand the impact of social adjustment among individuals with chronic pain."[8]

Meanwhile, the prescription drug landscape has been forever altered by the opioid epidemic. Not that these medications have gone away—opioids are still an effective barrier to short-term acute pain, and that's not likely to change—but physicians are much less likely to prescribe them for long-term use.

Here are two examples of potential prospects, albeit with some "yes, but . . ." disclaimers:

"Ziconotide is an analgesic that was approved by the US FDA in 2004 for the management of severe chronic pain in patients who are intolerant of or not responding to other treatments. It's injected into the spinal cord to block the transmission of pain signals. The evidence produced so far shows that ziconotide might be effective for these patients. It was, however, also found to be associated with some adverse events, including abnormal gait, dizziness, rapid involuntary eye movements, confusion, urinary retention, nausea, and vomiting.

"Tanezumab, a treatment for osteoarthritis-related pain or chronic low back pain, is said to target and stop the nerve growth factor activity to reduce pain signals. The drug was previously studied as an intravenous infusion but is now being studied as a subcutaneous injection. Tanezumab has not yet been approved by the FDA—in 2010, the FDA put a hold on clinical trials of the drug when some participants developed rapidly progressive osteoarthritis and osteonecrosis that led to the need for joint replacements. However, the hold has since been lifted, and tanezumab has now been granted Fast Track designation for an expedited review. The evidence from studies conducted so far suggests that tanezumab may be effective for reducing pain related to both indications, with the main adverse effects reported to be abnormal peripheral sensations (e.g., numbness and tingling), especially at higher doses, and, in the one study in patients with chronic low back pain, joint pain, pain in the extremities, and headache. It also appeared to result in a small number of patients requiring joint replacements in the osteoarthritis-related studies."[9]

Or, from another angle:

"Whereas Nav1.7 is expressed in all pain-sensing neurons, different TRP channels (or combinations of them) are found in subsets of pain-sensing neurons, making each subset sensitive to a specific temperature range. TRP channels comprise a 'superfamily' of related proteins that are found throughout the animal kingdom, where they serve a range of functions. Vampire bats have evolved the ability to use TRPV1 to detect infrared heat signatures in order to bite a blood vessel. Pit vipers, boas, and pythons have the most sensitive of all known TRPV1 channels, which enables the infrared vision that helps them hunt small, furry creatures at night.

"In humans, TRP channels are known to play important roles in neuropathic and inflammatory pain, and mutations in one of them are associated with a rare inherited disorder called familial pain syndrome.

"The discoveries of these mutations prompted a search for new painkillers, which has been slightly more successful than the experimental drugs targeting Nav1.7, but still has not produced any approved drugs. A number of small-molecule drugs that modulate TRP channels have been identified, and given the diverse roles of TRP channels in the body, drug discovery efforts targeting these channels have recently been expanded beyond pain management into other disease areas, including asthma, cancer, and metabolic disorders. But progress is slow, with few candidates making it to clinical trials, and none as yet approved. A better understanding of TRP channel biology will likely be needed to accelerate the discovery of any new drug treatment."[10]

Centuries after Galen's theriac, we're talking about pit vipers again.

The opioid epidemic has also focused more attention on alternative methods of treating chronic pain, some of them centuries old.

Having grown up in Eastern culture, I became familiar from childhood with yoga, tai chi, and other concepts that are only now taking hold in the Western world under the "new age" umbrella. Moreover, I have been told by more than a few chronic pain patients that some of these things have indeed helped to moderate their discomfort.

Chronic pain is often exacerbated by depression and stress, and most of the alternative approaches to managing that pain involves some form of relaxation therapy. In the case of tai chi and yoga, enticing the body to move—just as in physical therapy—is often helpful, both on a mental and muscular level.

Our charge as physicians is "do no harm," and I try to apply that philosophy to alternative chronic pain therapies. I have no problem with a patient using them in conjunction with scientific, evidence-based forms of healing as long as there is no potential for worsening the condition. For example, I have become intrigued with approaches like cognitive behavioral therapy and biofeedback as a possible weapon against pain.

On the other side of the fence, however, is a natural skepticism. Such practices as acupuncture and chiropractic are viewed with a certain amount of suspicion by the medical community as a whole because they are not evidence-based and have not proven effective by what traditional medicine views as scientific analysis. Most traditionally trained physicians like me admit that they share that skepticism.

Additionally, we as medical doctors are rigorously trained for years—medical school, residency, and fellowship training—to be able to locate, diagnose, and manage that pain, knowledge that a person administering an alternative option does not possess.

"Nerve blocks are most often used in my field for lower back pain, RSD/CRPs, fibromyalgia, migraines, and neuralgias. They have obvious advantages over drug treatment, although the two are often employed in tandem. There are fewer chances of drug-related side effects, no possibility of addiction, and a much faster result. When injected, medication can be hyper-focused on the affected areas, as opposed to being diffused throughout the body when they are taken systemically."[11]

Even if the results of a nerve block may not be permanent—and few are—giving a patient a respite from pain can be beneficial beyond the obvious. Besides easing their hurt, these may provide a window of time for them to begin (or step up) physical therapy to achieve a wider range of motion. This, in turn, may have long-lasting effects of its own.[12]

The neuromodulation techniques including peripheral nerve, spinal cord, and brain stimulation have major advantages as compared to nerve blocks in selected cases. Unlike a finite amount of local anesthetic, steroids, or other pain-killing drugs, the electrical modulation the stimulation creates is long-lasting, consistent, and—perhaps best of all—controlled by its owner, much like a morphine pump but without the serious risks of opioids.

The procedure is relatively new, having first been performed as a reaction to the "gate control" theory. An article on the website *Gizmag* reports: "The first use of small electrical currents as a human analgesic was reported in 1971—a Japanese team found that by sending small electrical currents through wires running along the epidural space—the outermost part of the spinal canal—they could disrupt the crippling pain signals being sent back to the brain, and replace them with a much more pleasant buzzing sensation.

"Spinal Cord Stimulation (SCS) is not a cure for chronic pain conditions—it's just a pain relief strategy. But it can be extremely effective, reducing the pain by 50 percent or more without the side effects associated with opioid analgesics like morphine. It's viewed as somewhat of a last resort for cases in which patients are unable to achieve a satisfactory quality of life using other pain relief methods."[13]

In the meantime, a wide range of technology-related panaceas are being revisited. Use of traditional methods like transcutaneous electrical nerve stimulation (TENS) has been in practice for many decades now and continues to provide meaningful relief in many cases.

"A recent phase 2 clinical trial provides the strongest evidence yet that a technique called transcutaneous electrical nerve stimulation (TENS) can alleviate pain for people with fibromyalgia, a complex, long-term condition characterized by widespread musculoskeletal pain accompanied by fatigue. Analgesics are only modestly effective in treating fibromyalgia, and clinicians are increasingly recognizing that non-pharmacological means like TENS are likely to be more effective and should be considered first.

"The trial, carried out by researchers at the University of Iowa and Vanderbilt University Medical Center, recruited 300 women diagnosed with fibromyalgia and randomly divided them into three groups. Members of one group used a TENS device at home for two hours a day for four weeks, alongside their regular treatment of light exercise. (The small battery-operated device delivers mild electrical currents via electrode pads attached directly to the skin.) Members of the second group used a sham device that delivered milder, non-therapeutic currents for the same period, while members of the third group continued with just their regular treatment. Before, during, and after the intervention, all of the participants filled out questionnaires designed to gauge their perceived pain.

"The results are encouraging. Those who had received TENS reported a significant reduction in both movement-related pain and fatigue. Almost half of them reported at least a 30 percent reduction in pain, compared to less than one quarter of those who received sham TENS, and just 14 percent of those in the third group, who did exercise alone. As for side effects, fewer than 1 in 20 of the participants who received TENS reported experiencing any pain or irritation from the electrodes."[14]

Ultrasound is also being reimagined for use in chronic pain treatment. Initially, it was primarily employed as a tool to ease pain, similar to a high-tech massage. Now, though, researchers believe they have found a way to use low-intensity ultrasound energy called transcranial focused ultrasound to actually intervene in some of the brain's microscopic conversations.

"A team led by Bin He, Ph.D., professor of biomedical engineering at Carnegie Mellon University, and funded in part by the National Institute of Biomedical Imaging and Bioengineering (NIBIB), has demonstrated the potential of a neuromodulation approach that uses low-intensity ultrasound energy, called transcranial focused ultrasound, or tFUS. In a paper published in the May 4, 2021, issue of *Nature Communications*, the authors describe tFUS in experiments with rodents that demonstrate the non-invasive neuromodulation alternative.

"For their studies, He and his team designed an assembly that included an ultrasound transducer and a device that records data from neuron signals, called a multi-electrode array. During experiments with anesthetized rodents, the researchers penetrated the skull and brain with various brief pulses of acoustic waves, targeting specific neurons in the brain cortex. They simultaneously recorded the change in electrophysiological signals from different neuron types with the multi-electrode array.

"When the researchers used tFUS to emit repeated bursts of ultrasound stimulation directly at excitatory neurons, they observed an elevated impulse rate or spike. They observed that inhibitory neurons subjected to the same tFUS energy did not display a significant spike rate disturbance. The study demonstrated that the ultrasound signal can be transmitted through the skull to selectively activate specific neuron sub-populations, in effect targeting neurons with different functions."[15]

Space and time do not enable me to compile an exhaustive list of drugs, devices, and other emerging weapons in the fight against chronic pain. Fortunately, there is Google.

Such a list would not be complete, however, without some reference to current research into the effects of heredity and genes on chronic pain. As Steven H. Richeimer and John J. Lee reported in *Practical Pain Management*:

"Currently, the largest area of research in pain medicine and genetics has been in the field of drug metabolism. Individuals all process and metabolize drugs to differing degrees. While the practical effect of this variation can be as minimal as requiring a slightly larger dose of gabapentin, it can be quite profound when a patient has nontherapeutic levels of anticoagulation because of the way his or her body metabolizes warfarin.

"A prime example of this in pain medicine is the metabolism of codeine. Codeine is metabolized by hepatic enzymes, specifically the CYP2D6 enzyme. Codeine itself is not analgesic; it becomes effective in treating pain only when it is metabolized into morphine. Some individuals completely lack the alleles needed to produce functioning CYP2D6, and they are considered poor metabolizers. These poor metabolizers may never achieve pain relief since they cannot form the active metabolite morphine. By contrast, individuals with multiple copies of CYP2D6, ultrarapid metabolizers, can have fatally high levels of the drug on standard doses. Other opioids that are partially dependent on CYP2D6 for metabolism include hydrocodone, meperidine, methadone, oxycodone, and tramadol.

"The focus of genetics in pain medicine pertains not only to medications but also to predicting patient response to pain. It is clear that not all individuals develop chronic pain from equal insults. For example, not all patients with shingles develop postherpetic neuralgia, and not all patients with diabetes develop painful peripheral neuropathy. Although it is difficult to control for cultural and environmental factors that may influence a patient's response to pain, there seems to be a genetic component to an individual's response to pain or the likelihood of developing chronic pain."[16]

With pain research, at least, the trend appears to be shifting toward "individualized" medicine. After centuries of operating under the assumption that all humans are alike and thus respond similarly to the same medications and treatment methods, it has now become apparent that such things as pain threshold and addiction potential may be as diverse as our genes.

Maybe that's the wild card.

Selected Bibliography

Chapter 1: The Mystery

Alpert, Joseph S. "Practicing Medicine in Plato's Cave." *The American Journal of Medicine* 117, no. 6 (June 2006): 455–56.

Astyrakai, Elisabeth. "References to Anaesthesia, Pain and Analgesia in the Hippocratic Collection." *Anesthesia and Analgesia* 110, no. 1 (January 2010): 188–94.

Baluyamu, Stavros J. "Galen and the Neuroscience." *Journal of Neurology and Stroke* 4, no. 1 (January 11).

Benini, Arnaldo. "Rene Descartes' Philosophy of Pain." *Spine* 24, no. 20 (October 15): 2115.

Burmistr, Iana. "Theories of Pain up to Descartes and after Neuromatrix." *Pain Medicine Journal* 3, no. 1 (2018).

Cheng, Wei. "A Battle against Pain? Aristotle, Theophrastus and the Physiologoi in Aspasius, On Nicomachean Ethics 156.14–20." *Phronesis* 62, no. 4 (2017): 392–416.

Curzer, Howard. "Aristotle's Painful Path to Virtue." *Journal of the History of Philosophy* 40, no. 2 (April 2002): 141–62.

Destree, Pierre. "Aristotle on the Paradox of Tragic Pleasure." In *Suffering Art Gladly: The Paradox of Negative Emotion in Art*, edited by Jerrold Levinson, 3–27. London: Palgrave-Macmillan, 2014.

Duncan, Grant. "Mind-Body Dualism and the Biopsychological Model of Pain: What Did Descartes Really Say?" *The Journal of Medical History* 25, no. 4 (2000): 495–513.

Jiminez, Marta. "Aristotle on Steering the Young by Pleasure and Pain." *The Journal of Speculative Philosophy* 29, no. 2: 137–64.

Jouanna, Jacques, and Neil Allies. *Greek Medicine from Hippocrates to Galen: Selected Papers*. Edited by Philip van der Eijk. Leiden: Brill, 2012.

Kleisiaris, Christos. "Health Care Practices in Ancient Greece: The Hippocrates Ideal." *Journal of Medical Ethics and History of Medicine* 7, no. 6 (March 2014).

Kristijansson, Kristjan. "Reason and Intuition in Aristotle's Moral Philosophy." *Journal of the History of Philosophy* (November 2020).
Mongiedi, Massieh. "Theories of Pain from Specificity to Gate Control." *Journal of Neuropsychology* (January 2013).
Scullin, Sarah, "Hippocratic Pain." PhD diss., University of Pennsylvania, 2012. *Publicly Accessible Penn Dissertations*. https://repository.upenn.edu/edissertations/578/.
Tashanly, Osama, and Mark I. Johnson. "Avicenna's Concept of Pain." *Libyan Journal of Medicine* (September 2010).
Uddin, Zakir. "Understanding Pain and Suffering from the Core Concepts of Descartes and Spinoza." *The Canadian Physiotherapy Association* (n.d.): https://physiotherapy.ca/understanding-pain-and-suffering-core-concept-descartes-and-spinoza.

Chapter 2: The Body's CEO

Ahmady, Asma Hayati. "The Brain in Pain." *Malaysian Journal of Medical Science* (December 2014).
Baliki, Marwan. "Beyond Feeling: Chronic Pain Hurts the Brain, Disrupting the Default Mode." *Journal of Neuroscience* (February 2008).
Batista-Garcia-Ramo, Karla, and Ivette Fernandez-Verdecia. "The Brain Structure: Function, Relationhip and Caridad." *Behavioral Sciences* 8, no. 4 (April 2018).
Kim, Dongwon, Younbyoung Chae, Hi-Joon Park, and In-Seon Lee. "Effects of Chronic Pain Treatment on Altered Functioning and Metabolic Activities in the Brain." *Fronters in Neuroscience* (July 2021).
Lindsay, Grace W. "Attention in Psychology, Neuroscience and Machine Learning." *Computer Neuroscience* (April 2020).
Messe, Arnaud, David Rudrauf, Habib Benali, and Guillaume Marrelec. "Relating Structure and Function in the Human Brain: Relative Contributions of Anatomy, Stationary Dynamics, and Non-stationarities." *PLOS Computational Biology* (March 2014).
Xu, Anna, Bart Larsen, Alina Henn, Erica B. Baller, J. Cobb Scott, Vaishnavi Sharma, Azeez Adebimpe, Allan I. Basbaum, Gregory Corder, Robert H. Dworkin, Robert R. Edwards, Clifford J. Woolf, Simon B. Eickhoff, Claudia R. Eickhoff, and Theodore D. Satterthwaite. "Brain Responses to Noxious Stimuli in Patients with Chronic Pain." *JAMA Network* (January 2021).

Chapter 3: Pain in Our Culture

Absolmzhavejad, Ali, Mohsen Banaie, Nader Tavakoli, Mohammad Safdari, and Ali Rajabpour-Sanati. "Pain Management in the Emergency Department." *Advanced Journal of Emergency Medicine* 2, no. 4 (2018).

Bernard, Stephen A. "Management of Pain in the United States—A Brief History and Implications for the Opioid Epidemic." *Health Series Insights* (December 2018).

Fillingin, Roger. "Individual Differences in Pain: Understanding the Mosaic That Makes Pain Personal." *HHS Public Access* (April 2017).

———. "Sex, Gender and Pain." *The Journal of Pain* 10, no. 5 (May 2009): 447–85.

Sinatra, Raymond. "Causes and Consequences of Inadequate Management of Acute Pain." *Pain Medicine* 11, no. 12 (December 2010).

Trachsel, Lindsay A., and Marco Cascella. *Pain Theory*. Treasure Island, FL: StatPearls, 2021.

Chapter 4: All the Hurtful Things

Bueno-Gomez, Noela. "Conceptualizing Suffering in Pain Philosophy." *Ethics and Humanity in Medicine* 12, no. 7 (September 2017).

Cetkin, Murat, Gulbin Ergin, Golgem Mehmetoglu, Beraat Alptug, Ayse Kilic, and Necati Ozler. "The Pain in the Canon of Medicine: Types, Causes and Treatments." *European Journal of Therapeutics* 25, no. 3 (September 2019).

Dahlhamer, James, Jacqueline Lucas, Carla Zelaya, Richard Nahin, Sean Mackey, Lynn DeBar, Robert Kerns, Michael Von Korff, Linda Porter, and Charles Helmick. "Prevalence of Chronic Pain and Higher Impact Chronic Pain Among Adults." *Morbidity and Mortality Weekly Report* 67, no. 36 (September 2018).

Payne, Richard. "Challenges in the Assessment and Management of Cancer Pain." *Journal of Pain and Symptom Management* 18, no. 1 (January 2000): 12–15.

Ponnalyan, Deepa, K. M. Bhat, G. S. Bhat, and D. J. Victor. "Oro Facial Pain: An Overview." *International Journal of Current Research and Review* 4, no. 9 (May 2012).

Raffaedi, William, and Elisa Arnando. "Pain as a Disease: An Overview." *The Journal of Pain Research* (August 2017).

Richard, Julianne Y., Robin Hurley, and Katherine H. Tabor. "Fibromyalgia: Centralized Pain Processing and Neurology." *The Journal of Neuropsychology* (July 2019).

Tabor, Abby, Michael A. Thacker, G. Lorimer Moseley, and Konrad P. Kording. "Pain: A Statistical Account." *Computational Biology* (January 2017).

Vivekanthom, A., Claire Edwin, Tamar Pincus, Manjit Matharu, Helen Parsons, and Martin Underwood. "The Association between Headache and Low Back Pain: A Systematic Review." *The Journal of Headache and Pain* (July 2019).

Watson, Jane C., and Paola Sandroni. "Central Neuroplathic Pain Syndrome." *Symposium on Pain Medicine* 91, no. 3 (March 2016): 372–85.

Chapter 5: Chasing Relief

Dublin, Adrienne, and Adam Patapouton. "Nociceptors: The Sensors of the Pain Pathway." *Journal of Clinical Investigation* (November 2010): 2522–25.

Schumann, Martin F. "Treatment Effectiveness and Medication Use Reduction for Older Adults in Interdisciplinary Pain Rehabilitation." *Mayo Clinic Proceedings* 4, no. 1 (June 2020): 276–86.

Chapter 6: How Chronic Pain Changed the Game

Addison, Robert. "Chronic Pain Syndrome." *The American Journal of Medicine* 77, no. 3 (September 1984): 54–58.

Basbaum, Allan. "Chemogenetic Management of Neuropathic Pain." *Brain* 140, no. 10 (October 2017): 3760–72.

Benedelon, Gillian. "Chronic Pain Patients and the Biomedical Model of Pain." *AMA Journal of Ethics* (May 2013).

Feher, Gergeley. "Chronic Pain Hurts the Brain: The Pain Physician's Perspective." *Behavioral Neurology* (2020).

Katz, M. "The Impact of Pain Management on Quality of Life." *Journal of Pain and Symptom Management* 24, no. 1 (July 2002).

Lima, Daniela Danks, Vera Lucia Pereira Alves, and Egberto Ribeiro Turato. "The Phenomenological-Existential Comprehensive of Chronic Pain." *Philosophy, Ethics and Humanities* (April 2014).

Mills, Sarah E. "Chronic Pain: A Review of Its Epidemiology and Associated Factors in Population-Based Studies." *British Journal of Anesthesia* (August 2019): 273–83.

Saldecker, Saquin. "Practical Approaches to a Patient with Chronic Pain of Unknown Etiology in Primary Care." *Journal of Pain Research* (September 2019).

Seth, Barta, and Lorraine de Gray. "Genesis of Chronic Pain." *Anesthesia and Intensive Care Medicine* 17, no. 9 (September 2014).

Yang, Seoyon, and Min Cheol Chang. "Chronic Pain: Structural and Functional Changes in Brain Structures and Associated Negative Affective States." *International Journal of Molecular Sciences* 20, no. 13.

Chapter 7: Beware the Evil Twin

Behr, Emily. "Chasing the Pain Relief, Not the High." *PLOS One* (March 2020).

DuPont, Robert. "The Opioid Epidemic Is an Historic Opportunity to Improve Both Prevention and Treatment." *Brain Research Bulletin* 138 (April 2018): 112–14.

Glod, Susan A. "The Other Victims of the Opioid Epidemic." *New England Journal of Medicine* (June 2017): 2101–2.

Haley, Danielle F. "The Opioid Epidemic during the COVID-19 Pandemic." *JAMA* (September 2020).

Hansen, Helena, and Julie Netherland. "Is the Prescription Opioid Epidemic a White Problem?" *American Journal of Public Health* (December 2016).

Jones, Mark R., Omar Viswanath, Jacquelin Peck, Alan D. Kaye, Jatinder S. Gill, and Thomas T. Simopoulos. "A Brief History of the Opioid Epidemic and Strategies for Pain Medicine." *Pain and Therapy* (2018).

Lyden, Jennifer, and Ingrid A. Binswanger. "The United States Opioid Epidemic." *Seminars in Perinatology* 47, no. 3 (April 2019): 123–31.

Meldrum, Marcia. "The Ongoing Opioid Prescription Epidemic: Historical Context." *American Journal of Public Health* 106, no. 8 (August 2016): 1365–66.

Murphy, Livela. "Ending the Opioid Epidemic—A Call to Action." *New England Journal of Medicine* (December 2016): 2412–15.

Santoro, Taylor N., and Jonathan D. Jenkins. "Racial Bias in the U.S. Opioid Epidemic." *Cureus* 10 (December 2018).

Shipton, Edward, Elspeth Shipton, and Ashleigh J. Shipton. "A Review of the Opioid Epidemic: What Do We Do about It?" *Pain and Therapy* (July 18): 23–26.

Vadivelu, Nalini, Alice M. Kai, Vijay Kodumudi, Julie Sramcik, and Alan D Kaye. "The Opioid Crisis: A Comprehensive Overview." *Current Pain and Headache Report* (February 2018).

Wenger, Sarah, Jason Drott, Rebecca Fillipo, Alyssa Findlay, Amanda Genung, Jessica Heiden, and Joke Bradt. "Reducing Opioid Use for Patients with Chronic Pain." *Physical Therapy* 98, no. 5 (May 2018): 424–33.

Chapter 8: Anomalies

Channe, Stuart L. "Religion and the Theory of Masochism." *Journal of Religion and Health* (September 1983): 226–33.

Rosemberg, Sergey. "Congenital Insensitivity to Pain with Anhydrous (Heredity) and Autonomic Neuropathy." *Pediatric Neurology* 11, no. 1 (July 1994): 50–56.

Van der Kulk, Bessel A. "The Compulsion to Repeat the Trauma." *Psychiatric Clinics of North America* 12, no. 2 (June 1989): 389–411.

Chapter 9: Keeping Pain in Mind

Craig, K. D. "Psychology of Pain." *Postgraduate Medical Journal* (December 1984): 835–40.

Davey, Elizabeth S. "Psychology and Chronic Pain." *Pain* 17, no. 11 (2016): 568–70.

Flink, Ida K. "Pain Psychology in the 21st Century." *Scandanavian Journal of Pain* (April 2020).

Jacoby, Marilyn S. "Psychological Factors Influencing Chronic Pain and the Impact of Litigation." *Current Physical and Rehabilitation Report* 1 (2013): 135–41.

Linton, Stephen J., and William S. Shaw. "Impact of Psychological Factors in the Experience of Pain." *Physical Therapy and Rehabilitation Journal* 91, no. 5 (May 2011): 700–711.

Schatzberg, Alan S. "The Relationship of Chronic Pain and Depression." *Journal of Clinical Psychology* (2004).

Chapter 10: The Road Ahead

Clipper, Stephanie E. "What Is the Future of Pain Research?" SpineUniverse. March 22, 2016. www.spineuniverse.com/treatments/medication/what-future-pain-research.

Cohen, Steven P., Lene Vase, and William Hooten. "Chronic Pain: An Update on Burden, Best Practices and New Advances." *The Lancet* 397, no. 10289 (May 2021): 2082–97.

Collins, Sonya. "Beyond Opioids: The Future of Pain Management." *WebMD*. Last modified March 14, 2018. www.webmd.com/special-reports/opioids-pain/20180314/opioid-alternatives.

Gilron, Ian. "Current Status and Future Direction of Pain Related Outcome Measures for Post Surgical Pain Trials." *Canadian Journal of Pain* 3, no. 2 (2019).

Kress, Hans-George, Dominic Aldington, Eli Alon, Stefano Coaccioli, Beverly Collett, Flaminia Coluzzi, Frank Huygen, Wolfgang Jaksch, Eija Kalso, Magdalena Kocot-Kępska, Ana Cristina Mangas, Cesar Margarit Ferri, Philippe Mavrocordatos, Bart Morlion, Gerhard Müller-Schwefe, Andrew Nicolaou, Concepción Pérez Hernández, and Patrick Sichère. "A Holistic Approach to Chronic Pain Management That Involves All Stakeholders." *Current Medical Research and Opinion* 31, no. 9 (2015).

Macey, Jianren. "Current Challenges in Translational Pain Research." *Science* 33, no. 11 (November 2017): 568–93.

Roy, Ruben. "Neurofeedback for Pain Management: A Systemic Review." *Frontiers of Neuroscience* (July 2020).

Stern, Peter, and Leslie Roberts. "The Future of Pain Research." *Science* 354, no. 632 (November 2016).

Zappaterro, Mauro. "Future Research in Pain." *Pain Care Essentials and Innovation* (2021): 258–67.

Notes

Preface

1. Peter Abaci, *Take Charge of Your Chronic Pain* (Globe-Pequot Press, 2009).
2. Benjamin Schneider, "Fentanyl: The City's Biggest Public Health Crisis," *SF Weekly*, August 6, 2021.

Chapter 1: The Mystery

1. Adrianne Carley, "Pain: A Historical Perspective of Pain Theories," *Pathways* (November 2019).
2. Jamie Dow, "Aristotle's Theory of the Emotions: Emotions as Pleasures and Pains," Oxford Scholarship Online, Oxford University Press, 2011, https://oxford.universitypressscholarship.com/view/10.1093/acprof:oso/9780199546541.001.0001/acprof-9780199546541-chapter-3#:~:text=In%20his%20theory%2C%20to%20have,for%20the%20particular%20emotion%20experienced.
3. "Ancient, Medieval and Renaissance Theories of the Emotions," Stanford Encyclopedia of Philosophy.
4. David Conan Wolfsdorf, "Plato on Pain," *Antiquorum Philosophia: An International Journal* 9 (2015).
5. Elizabeth Astyrakaki, Alexandra Papaioannou, and Helen Askitopolo, "References to Anesthesia, Pain and Analgesia in the Hippocratic Collection," *Anasthesia and Analgesia* 10, 1 (January 2021): 188–94.
6. Astyrakaki, Papaioannou, and Askitopolo, "References to Anesthesia."
7. *The Genuine Works of Hippocrates*, trans. and ed. Charles Darwin Adams (New York: Dover, 1868).
8. Noel Si-Yang Bay and Boon-huat Bay, "Greek Anatomist Herophilus: Father of Anatomy," *Anatomy and Cell Biology* (2010): 280–83.
9. Bay and Bay, "Greek Anatomist Herophilus."

10. Courtney Roby, "Galen on the Patient's Role in Pain Diagnosis: Sensation, Consensus, and Metaphor," *Studies in Ancient Medicine* 45 (2016): 304–22.
11. Roby, "Galen."
12. Roby, "Galen."
13. Roby, "Galen."
14. Stavros Balloyanis Jr., "Galen and the Neurosciences," *Journal of Neurology and Stroke* (January 2016).
15. Roby, "Galen."
16. Roby, "Galen."
17. Kremer Aschill, "The Scientific History of Hydrocephalus and Its Treatment," *Neurological Review* 22 (1999): 68–93.
18. Aschill, "The Scientific History of Hydrocephalus."
19. Kern Olson, "A History of Pain," *Practical Pain Management* 13, 6 (July 2013).
20. Anne Abbott, "The Man Who Bared the Brain," *Nature* (2015).
21. "Philosophy," Western Civilization II Guides, December 14, 2013, https://westerncivguides.umwblogs.org/2013/12/14/philosophy/.
22. Olson, "A History of Pain."
23. Arnaldo Benini, "Rene Descartes' Philosophy of Pain," *Spine* 24, no. 20 (October 15): 2115.
24. Massieh Moayedi and Karen D. Davis, "From Specificity to Gate Control," *Journal of Neurophysiology* (January 2013).
25. Moayedi and Davis, "From Specificity to Gate Control."
26. "History of Neuroscience: Charles Scott Sherrington," Neuroscientifically Challenged, May 31, 2017, https://neuroscientificallychallenged.com/posts/history-of-neuroscience-charles-scott-sherrington.
27. Award Ceremony Speech, NobelPrize.org, December 10, 1932, www.nobelprize.org/prizes/medicine/1932/ceremony-speech/.

Chapter 2: The Body's CEO

1. "Anatomy of the Brain," American Association of Neurological Surgeons, www.aans.org/Patients/Neurosurgical-Conditions-and-Treatments/Anatomy-of-the-Brain.
2. "A Conversation with Ray Kurzweil," World Knowledge Forum, November 2020, www.wkforum.org/WKF/2020/en/wnf_view.php?no=2477&search_year=2020.
3. "A Conversation with Ray Kurzweil."
4. "A Conversation with Ray Kurzweil."
5. A. A. Fanous and W. T. Couldwell, "Transnasal Excerebration Surgery in Ancient Egypt," *Journal of Neurosurgery* (2012): 743–48.
6. D. S. Lamb, "Mummification, Especially of the Brain," *American Anthropologist* 3, no. 2 (1901): 294–304.

7. Bill Bryson, *The Body: A Guide for Occupants* (New York: Anchor, 2019).
8. Antonio Damasio, "The Quest to Understand Consciousness," TED talk, December 12, 2011.
9. Tereza Pultarova, "Why Human Head Transplants Will Never Work," LiveScience, November 20, 2017.
10. Arthur Caplan, "Doctor Seeking to Perform Head Transplants Is Out of His Mind," *Forbes*, February 21, 2015.
11. Bryson, *The Body*.
12. *Rain Man*, directed by Barry Levinson (1988).
13. Michael Greshko, "Human Memory: How We Make, Remember and Forget Memories," *National Geographic*, March 4, 2019.
14. Linda Rodriguez McRobbie, "Total Recall: The People Who Never Forget," *Guardian*, February 8, 2017.
15. Yvaine Ye, "Photographic Memory," *New Scientist*, www.newscientist.com/definition/photographic-memory/.
16. Michael A. Yassa, "Hippocampus," Encyclopedia Brittanica, October 30, 2020, https://www.britannica.com/science/hippocampus.
17. Randy Rieland, "Pain and the Brain," *Smithsonian Magazine*, February 24, 2012, www.smithsonianmag.com/innovation/pain-and-the-brain-107370345/.
18. Rieland, "Pain and the Brain."
19. Rieland, "Pain and the Brain."
20. "41 Fascinating Facts about Your Amazing Brain," Ask the Scientists, https://askthescientists.com/brain-facts.
21. Eric Haseltine, "5 Impossible Things Your Brain Can Do," *Psychology Today*, June 20, 2018.
22. Haseltine, "5 Impossible Things Your Brain Can Do."
23. Alreza Minager, John Ragheb, and Roger E. Kelley, "The Edwin Smith Surgical Papyrus: Description and Analysis of the Earliest Case of Aphasia," *Journal of Medical Biography* (May 2003).
24. Nancy Kanwisher. "A Neural Portrait of the Human Mind," TED talk, October 1, 2014.
25. "Brain Basics," National Institutes of Health, www.ninds.nih.gov/health-information/patient-caregiver-education/brain-basics-know-your-brain.
26. "Understanding the Stress Response," Harvard Medical School, July 6, 2020, www.health.harvard.edu/staying-healthy/understanding-the-stress-response.
27. Choong-Wan Woo, Mathieu Roy, Jason T. Buhle, and Tor D. Wager, "Distinct Brain Systems Mediate the Effects of Nociceptive Input and Self-Regulation on Pain," *PLoS Biology* 13, no. 1 (January 2015).
28. Glynis Sherwood, "Going the Distance: Lessons from Marathon Runners in Overcoming Chronic Emotional Pain," Online Psychology, August 17, 2012.

29. David Cosio, "The Perseverance Loop: The Psychology of Pain and Factors in Pain Perception," *Practical Pain Management* 20, 1 (2020).
30. "Gate Control Theory," Wikipedia, April 28, 2022, https://en.wikipedia.org/wiki/Gate_control_theory.
31. Ronald Melzack, "Pain: An Overview," *Acta Anaesthesiologica Scandinavica* 43, no. 9 (October 1999): 880–84.
32. Aaron Dertel, "Tributes Pour in for Pain Research Pioneer Ronald Melzack," *Montreal Gazette*, January 7, 2020.
33. Dertel, "Tributes Pour In."

CHAPTER 3: PAIN IN OUR CULTURE

1. T. S. Kelly, "OTC Analgesic Makers' Ads Seek Pain Relief from Private Label," Drug Store News, June 11, 2018, https://drugstorenews.com/insights/otc-analgesic-makers-ads-seek-pain-relief-from-private-label.
2. Sagar Mukhekar, Sushant Terdale, and Onkar Sumant, "Opportunity Analysis and Industry Forecast," *U.S. Topical Pain Relief Market, 2020–2027*, www.alliedmarketresearch.com/us-topical-pain-relief-market#:~:text=The%20U.S.%20topical%20pain%20relief,area%20of%20inflammation%20or%20pain.
3. Mukhekar, Terdale, and Sumant, "Opportunity Analysis and Industry Forecast."
4. Erika Kinetz, "Fake Doctors, Misleading Claims Drive Oxycontin China Sales," Associated Press, November 20, 2019.
5. Rosemary Black, "Support Groups for Pain," *Practical Pain Management*, July 22, 2020, https://patient.practicalpainmanagement.com/resource-centers/chronic-pain-management-guide/support-groups-pain.
6. Kai Karos and Joanna McFarland, "The Social Threats of COVID-19 for People with Chronic Pain," *Pain Magazine* (October 2020): 2229–35.
7. Philomena Puntillo, "Best Practices and Research," *Clinical Anaesthesiology* 34, no. 3 (September 2020): 529–37.
8. *Daily Strength*, www.dailystrength.org.
9. Paul Muschick, "Relief Needed from Scam Offering Anti-Pain Cream," *Morning Call*, December 31, 2014, www.mcall.com/news/watchdog/mc-pain-relief-cream-phone-sale-scam-watchdog-20141231-column.html.
10. Dmitry Arbuck and Amber Fleming, "Pain Assessment: Review of Common Tools," *Practical Pain Management*, April 29, 2019, www.practicalpainmanagement.com/resource-centers/opioid-prescribing-monitoring/pain-assessment-review-current-tools.
11. Patti Neighmond, "Words Matter When Talking about Pain with Your Doctor," National Public Radio, July 29, 2018.
12. Neighmond, "Words Matter."
13. "Dolorimeter," Wikipedia.
14. Neighmond, "Words Matter."

15. "What Every Nurse Needs to Know about Pain Management." *Daily Nurse*, August 24, 2016.
16. Marcia Carteret, "Cultural Aspects of Pain Management," Dimensions of Culture, November 2, 2010, http://www.dimensionsofculture.com/2010/11/cultural-aspects-of-pain-management/.
17. Ozen Dedeli and Gulten Kaptan, "Spirituality and Religion in Pain and Pain Management," *Health Psychology News*, September 24, 2013.
18. David E. Weissman, Deborah Gordon, and Shiva Bidar-Sielaff, "Cultural Aspects of Pain," *Journal of Palliative Medicine* 7, no. 5 (October 2004).
19. Sarah M. Whitman, "Pain and Suffering as Viewed by the Hindu Religion," *The Journal of Pain* 8, no. 8 (August 2002): 667–73.
20. Whitman, "Pain and Suffering."
21. Darlene Cohen, "Mindfulness and Pain," www.darlenecohen.net/welcome/mindfulness.html.
22. Alcira Molina-Ali, "In Search of a Muslim Pain Principle," YouTube, May 7, 2010. Accessed September 14, 2022, https://www.youtube.com/watch?v=7aNcG-GeUl8.
23. Sue Peacock and Shilpa Patel, "Cultural Influences on Pain," International Association for the Study of Pain, March 2008, www.ncbi.nlm.nih.gov/pmc/articles/PMC4589930/.
24. Alyson Fincke, "Genetic Influences on Pain Perception and Treatment," *Practical Pain Management* 10, no. 1 (2011).
25. "Study Finds Link between Red Hair and Pain Threshold," NIH Research Matters, April 20, 2021.
26. R. B. Fillingin, "Genetic Contributions to Pain: A Review of Findings in Humans," *Oral Diseases* (November 2008): 673–82.
27. Nancy Wells, Chris Pasero, and Margo McCaffrey, "Improving the Quality of Care through Pain Assessment," in *Patient Safety and Quality: An Evidence-Based Handbook for Nurses* (Rockville, MD: Agency for Healthcare Research and Quality, 2008).
28. Michael Schatman, "The Role of the Insurance Industry in Perpetuating Suboptimal Pain Management," *Pain Medicine* 12, no. 3 (March 2011): 415–26.
29. Lynn Webster, "Pre-Existing Conditions Deserve Affordable Treatment," Pain News Network, October 3, 2020. Accessed September 14, 2022, https://www.painnewsnetwork.org/stories/2020/10/3/pre-existing-conditions-deserve-affordable-treatment.
30. Weissman, Gordon, and Bidar-Sielaff, "Cultural Aspects of Pain."
31. Nivet and Taylor, "Structural Racism in Pain Practice."
32. Janice A. Sabin, "How We Fail Black Patients in Pain," AAMC Insights, January 6, 2020, www.aamc.org/news-insights/how-we-fail-black-patients-pain.
33. Ronald Wyatt, "Pain and Ethnicity," *AMA Journal of Ethics* (May 2013).

34. Nicole F. Roberts, "Emotional and Physical Pain Are Almost the Same—To Your Brain," *Forbes*, February 14, 2020.

Chapter 4: All the Hurtful Things

1. Bill Bryson, *The Body: A Guide for Occupants* (New York: Anchor Publishing, 2019).
2. Nessa Coyle, "In Their Own Words: Seven Advanced Cancer Patients Describe Their Experience with Pain and the Use of Opioid Drugs," *Journal of Pain and Symptom Management* 27, no. 4 (April 2004): 300–309.
3. Jacqueline Andriakos, "What Does Appendicitis Pain Really Feel Like? 13 People Who Have Been There," *Self*, October 16, 2018.
4. "Kidney Stone Pain: Firsthand Recollections of the Experience," Intermountain Health Care, Patient Stories and Blog, September 2017, https://intermountainhealthcare.org/blogs/topics/live-well/2017/09/kidney-stone-pain-firsthand-recollections-of-the-experience/.
5. Wim Dekkers, "Hermeneutics and Experiences of the Body. The Case of Low Back Pain," *Theoretical Medicine and Bioethics* (1998): 277–93.
6. Clifford Woolf, "What Is This Thing Called Pain?" *The Journal of Clinical Investigation* (November 2010).
7. Les Barnsley, "Back Pain," in *The Musculoskeletal System*, ed. Philip Sambrook, Leslie Schrieber, Thomas Taylor, and Andrew M. Ellis (London: Churchill Livingstone, 2010), 247–59, https://doi.org/10.1016/B978-0-7020-3377-3.00004-4.
8. Richard Dawkins, *The Greatest Show on Earth* (New York: Free Press, 2009).
9. Bryson, *The Body*.
10. Bryson, *The Body*.
11. Michael Sinkin, *The Dental and the Incidental* (blog), https://michaelsinkindds.com/blog/page/3/.
12. "How to Locate the Source of Pain—Referred Pain," CBS News, January 31, 2002.
13. Walter Somerville, "A Pain in the Arm: A Symptom of Heart Disease," *The Postgraduate Medical Journal* (1959).
14. Paul Ingraham, "Why Does Pain Hurt?" PainScience.com, October 22, 2015.
15. Colin Barras, "The Real Reasons Why Childbirth Is So Painful and Dangerous." *BBC Earth*, December 22, 2016.
16. Barras, "The Real Reasons."
17. Barras, "The Real Reasons."
18. Barras, "The Real Reasons."
19. "Cancer Pain: Relief Is Possible," Mayo Clinic, February 12, 2021.
20. Allan Basbaum, "If the Brain Can't Feel Pain, Why Do We Get Headaches?" BrainFacts.com, September 24, 2014.

21. Karen Bannan, "New Drugs That Block a Brain Chemical Are Game Changers for Some Migraine Sufferers," *Science News*, March 22, 2021.
22. Margaret Twitty, "Learning More about CRPS/RSD," My Journey With CRPS/RSD, October 17, 2019, https://myjourneywithrsd.com/learning-more-about-crpsrsd.
23. Richard W. Hanson, "The Biological Aspects of Pain," *Self Management of Chronic Pain* (2006).
24. Forest Tennant, "Complications of Uncontrolled, Persistent Pain," *Practical Pain Management* 4, 1 (2012).
25. Tennant, "Complications."
26. Shaleek Blackburn, "How Pain Affects Your Mind and Body, Explained," *Press-Enterprise*, June 15, 2015.
27. "Neurogenic Shock," Wikipedia, last updated July 4, 2022, https://en.wikipedia.org/wiki/Neurogenic_shock.

Chapter 5: Chasing Relief

1. George Zaidan, "How Do Painkillers Work?" TED talk, January 4, 2018.
2. "Pain Theories," Wikipedia, last updated June 4, 2022, https://en.wikipedia.org/wiki/Pain_theories.
3. "Pain Theories," Wikipedia.
4. "Charles Scott Sherrington," *History of Neuroscience*, May 31, 2017.
5. Clifford J. Woolf, "Nociceptors: Noxious Stimulus Detectors," *Neuron* (July 2007).
6. Woolf, "Nociceptors."
7. Woolf, "Nociceptors."
8. McGill University, "Ascending Pain Pathways," *The Brain from Top to Bottom*, https://thebrain.mcgill.ca/flash/i/i_03/i_03_cl/i_03_cl_dou/i_03_cl_dou.html.
9. Bill Bryson, *The Body: A Guide for Occupants* (New York: Anchor, 2019).
10. Ben Spencer, "How Saying 'Ow!' Can Ease Your Pain," *Daily Mail*, February 1, 2015.
11. Judy Peakman, "A History of Opium," *History Today* 68, no. 10 (October 2018).
12. Demetros Karaberopoulos, Marianna Karamanou, and George Androutsos, "The Theriac in Antiquity," *Lancet* 379 (May 2012).
13. Karaberopoulos, Karamanou, and Androutsos, "The Theriac in Antiquity."
14. Karaberopoulos, Karamanou, and Androutsos, "The Theriac in Antiquity."
15. Peakman, "A History of Opium."
16. "Opium and Laudenum: History's Wonder Drugs," The Chemical Institute of Canada, July 2015.
17. Erick Trickey, "Inside the Story of America's 19th Century Opioid Addiction," *Smithsonian Magazine*, January 4, 2018.

18. Ananya Mandal, "Morphine History," *Medical Life Science News*, February 27, 2019, www.news-medical.net/health/Morphine-History.aspx.
19. Mandal, "Morphine History."
20. Trickey, "Inside the Story."
21. Mandal, "Morphine History."
22. Diarmuid Jeffreys, *Aspirin: The Remarkable Story of a Wonder Drug* (New York: Bloomsbury, 2008).
23. Elizabeth Landau, "From a Tree: A Miracle Called Aspirin," *CNN*, December 22, 2010.
24. "John R. Vane—Facts," NobelPrize.org, July 11, 2022, www.nobelprize.org/prizes/medicine/1982/vane/facts.
25. Landau, "From a Tree."
26. Zaidan, "How Do Painkillers Work?"
27. "Pharmaco-Kinetics and -Dynamics of Opioids," ATrain Education, www.atrainceu.com/content/4-pharmaco-kinetics-and-dynamics-opioids.
28. "Pharmaco-Kinetics and -Dynamics of Opioids," ATrain Education.

Chapter 6: How Chronic Pain Changed the Game

1. Melanie Thernstrom, *The Pain Chronicles* (New York: Farrar, Straus and Giroux, 2010).
2. Floyd E. Bloom, M. Flint Beal, and David J. Kupfer, *The Dana Guide to Brain Health* (New York: Dana Press, 2003).
3. "One in 3 Cancer Survivors Has Chronic Pain," American Cancer Society, June 28, 2019, www.cancer.org/latest-news/one-in-3-cancer-survivors-has-chronic-pain.html.
4. Jennifer Byrne, "Chronic Late Effects of Cancer Treatment: The Consequence of a Cure," *Healios* (April 2020).
5. Byrne, "Chronic Late Effects of Cancer Treatment."
6. Lenny Bernstein, "For Some with Chronic Pain, the Problem Is Not in Their Backs or Their Knees, but Their Brains," *Washington Post*, September 23, 2019.
7. Leslie J. Crofford, "Chronic Pain: Where the Body Meets the Brain," *Transactions of the American Clinical and Climatological Association* 126 (2015): 167–83.
8. Jeremy Gauntlett-Gilbert, Karen Rodham, Abbie Jordan, and Peter Brock, "Emergency Department Staff Attitudes toward People Presenting in Chronic Pain: A Qualitative Study," *Pain Medicine* (November 2015): 2065–74.
9. "Fibromyalgia," Wikipedia, accessed July 23, 2022, https://en.wikipedia.org/wiki/Fibromyalgia.
10. Hannah Nichols, "How to Recognize Fibromyalgia Flares," *Medical News Today*, March 20, 2018.

11. Akhtar Purvez, *Managing Pain in the Age of Addiction* (Lanham, MD: Rowman & Littlefield, 2018).
12. MDJunction Support Group, www.tapatalk.com/groups/mdjunctionextension/.
13. Bernstein, "For Some with Chronic Pain."
14. Purvez, *Managing Pain.*
15. Purvez, *Managing Pain.*
16. Purvez, *Managing Pain.*
17. Bernstein, "For Some with Chronic Pain."
18. "Acute to Chronic Pain Signatures," National Institutes of Health, January 19, 2022, https://commonfund.nih.gov/pain.
19. "Acute to Chronic Pain Signatures," National Institutes of Health.
20. Alix Spiegel, "Invisibilia: For Some Teens with Debilitating Pain, the Treatment Is More Pain," NPR, March 9, 2019, https://www.npr.org/sections/health-shots/2019/03/09/700823481/invisibilia-for-some-teens-with-debilitating-pain-the-treatment-is-more-pain.
21. Spiegel, "Invisibilia."
22. Laura Schocker, "More Than Just a Migraine: Battling Migraine Stigma," *Huffington Post*, July 16, 2011.
23. Schocker, "More Than Just a Migraine."
24. Bernstein, "For Some with Chronic Pain."
25. Linda Rath, "The Connection between Pain and Your Brain," Arthritis Foundation, www.arthritis.org/health-wellness/healthy-living/managing-pain/understanding-pain/pain-brain-connection.
26. "Explain the Pain: Is It Osteoathritis or Rheumatoid Arthritis?" Harvard Health Publishing, August 21, 2014, www.health.harvard.edu/pain/explain-the-pain--is-it-osteoarthritis-or-rheumatoid-arthritis.
27. "Low Back Pain," National Institutes of Health, March 2020, www.ninds.nih.gov/sites/default/files/migrate-documents/low_back_pain_20-ns-5161_march_2020_508c.pdf.
28. "Low Back Pain," National Institutes of Health."

Chapter 7: Beware the Evil Twin

1. Akhtar Purvez, *Managing Chronic Pain in an Age of Addiction* (Lanham, MD: Rowman & Littlefield, 2018).
2. Sarah DeWeerdt, "Tracing the U.S. Opioid Crisis to Its Roots," *Nature*, September 2019.
3. Nina Bai and Adam Smith, "Body's 'Natural Opioids' Affect Brain Cells Much Differently Than Morphine," *UCSF*, May 10, 2018, www.ucsf.edu/news/2018/05/410376/bodys-natural-opioids-affect-brain-cells-much-differently-morphine.
4. "Drug Overdoses," NSC Injury Facts, 2020, https://injuryfacts.nsc.org/home-and-community/safety-topics/drugoverdoses/data-details/.

5. Dina Fine Maron, "How Opioids Kill," *Scientific American*, January 2018.
6. Ronald Hirsch, "The Opioid Epidemic: It's Time to Place Blame Where It Belongs," *Journal of Missouri Medicine* (March 2017): 82–83.
7. "Opioid Overdose Crisis," National Institutes of Health, March 11, 2021, https://nida.nih.gov/research-topics/opioids/opioid-overdose-crisis.
8. "Timeline: Emergence of the Opioid Crisis," *CNN*, July 23, 2021.
9. DeWeerdt, "Tracing the U.S. Opioid Crisis."
10. Art Van Zee, "The Promotion and Marketing of OxyContin: Commercial Triumph, Public Health Tragedy," *American Journal of Public Health* (February 2009): 221–27.
11. DeWeerdt, "Tracing the U.S. Opioid Crisis."
12. "The More Opioids Doctors Prescribe, the More They Get Paid," Harvard School of Public Health, September 16, 2019, www.hsph.harvard.edu/news/hsph-in-the-news/opioids-doctors-prescriptions-payments/.
13. "Timeline: Emergence of the Opioid Crisis," *CNN*.
14. "Timeline: Emergence of the Opioid Crisis," *CNN*.
15. "Justice Department Announces Global Resolution of Criminal and Civil Investigations with Opioid Manufacturer Purdue Pharma and Civil Settlement with Members of the Sackler Family," US Department of Justice, Office of Public Affairs, October 21, 2020.
16. "Timeline: Emergence of the Opioid Crisis," *CNN*.
17. "Deadly Drugs and Targeting Youth: 6 Ways Mexican Drug Cartels Wreak Devastation in the US," *Louisville Courier Journal*, July 6, 2021, www.courier-journal.com/story/news/2021/06/30/mexican-cartels-5-ways-they-wreak-devastation-in-america/5248553001/.
18. "Prescription Drug Use Is a Risk Factor for Heroin Users," National Institute on Drug Abuse, January 2018, https://nida.nih.gov/publications/research-reports/prescription-opioids-heroin/prescription-opioid-use-risk-factor-heroin-use.
19. Sahoo Saddichha and Baxi Sinha, "The Role of Gateway Drugs and Psychological Factors in Substance Dependence in Eastern India," *International Journal of Psychiatry and Medicine* 37, no. 3 (2007): 257–66.
20. "Naloxone DrugFacts," National Institute on Drug Abuse, January 2022, https://nida.nih.gov/publications/drugfacts/naloxone.
21. "Naloxone DrugFacts," National Institute on Drug Abuse.
22. "Naloxone DrugFacts," National Institute on Drug Abuse.
23. Mark Tyndall and Zoë Dodd, "How Structural Violence, Prohibition, and Stigma Have Paralyzed North American Responses to Opioid Overdose," *AMA Journal of Ethics* 22, no. 8 (August 2020): 723–28.
24. Michael Schatman, "Identifying Abusers Prior to Initiating Chronic Opioid Therapy," *Practical Pain Management* 8, no. 1 (2008).
25. Schatman, "Identifying Abusers."

Chapter 8: Anomalies

1. David Cox, "The Curse of the People Who Don't Feel Pain," *BBC Future*, April 17, 2017.
2. Cox, "The Curse of the People Who Don't Feel Pain."
3. Jessie Higgins, "Real People: This Is Life with No Pain," *Evansville Courier and Press*, April 24, 2017.
4. Cox, "The Curse of the People Who Don't Feel Pain."
5. Cox, "The Curse of the People Who Don't Feel Pain."
6. "Congenital Insensitivity to Pain with Anhidrosis," Medline Plus, last updated May 1, 2011, https://medlineplus.gov/genetics/condition/congenital-insensitivity-to-pain-with-anhidrosis/.
7. Rachel Alter, "Feeling No Pain: What It's Like to Live with CIPA," *Brain World*, February 28, 2021.
8. Kamal Kant Kohli, "Man Who Feels No Pain," Medical Dialogue, March 23, 2019.
9. Katie Lambert, "How CIPA Works," HowStuffWorks.com, September 21, 2007, https://science.howstuffworks.com/life/inside-the-mind/human-brain/cipa.htm.
10. Stella Zhang, Saghira Malik Sharif, Ya-Chun Chen, Enza-Maria Valente, Mushtaq Ahmed, Eamonn Sheridan, Christopher Bennett, and Geoffrey Woods, "Clinical Features for Diagnosis and Management of Patients with PRD12 Congenital Insensitivity to Pain," *Journal of Medical Genetics* 52, no. 8 (August 2016).
11. Jennifer Welsh, "People Who Feel No Pain Can't Smell, Either," LiveScience, March 23, 2011.
12. Welsh, "People Who Feel No Pain Can't Smell, Either."
13. Welsh, "People Who Feel No Pain Can't Smell, Either."
14. Nicolas Danziger and J. C. Willer, "L'insensibilité congénitale à la douleur [Congenital insensitivity to pain]," *Revue neurologique* 165, no. 2 (2009): 129–36.
15. Cox, "The Curse of the People Who Don't Feel Pain."
16. Jessica Lear, "The Absence of Pain," *The Journal of Young Investigators* (July 2011).
17. Sophie Imhof, Tomislav Kokotović, and Vanja Nagy, "PRDM12: New Opportunity in Pain Research," *Trends in Molecular Medicine* 26, no. 12 (August 2020).
18. Cox, "The Curse of the People Who Don't Feel Pain."
19. Zaria Gorvett, "Why Pain Feels Good," *BBC Future*, October 1, 2015.
20. Gorvett, "Why Pain Feels Good."
21. The editors of Encyclopedia Britannica, "Masochism," Encyclopedia Britannica, last updated January 8, 2014, https://www.britannica.com/topic/masochism.

22. David J. Linden, "The Neurobiology of BDSM Sexual Practice," *Psychology Today*, March 20, 2015.
23. Kathryn Stangret and Kat Ramus, "Masochism," *Grey Matters*, February 2015.
24. Stangret and Ramus, "Masochism."
25. Albert Schrut, "A Psychodynamic (Non-Oedipal) and Brain Function Regarding a Type of Male Sexual Moschism," *Journal of the American Academy of Psychoanalysis and Dynamic Psychology* (February 2005).
26. "Sexual Masochism Disorder," *Psychology Today*, last updated September 15, 2021, www.psychologytoday.com/us/conditions/sexual-masochism-disorder.
27. Cara Dunkley, Craig Henshaw, Saira Henshaw, and Lori Brotto, "Pain or Pleasure, a Theoretical Perspective," *The Journal of Sex Research* (May 2019).
28. Dunkley, Henshaw, Henshaw, and Brotto, "Pain or Pleasure."
29. Dunkley, Henshaw, Henshaw, and Brotto, "Pain or Pleasure."
30. Daniel J. Winarick, "Masochism Explained: The Self-Sabotaging Personality," *Psychology Today*, June 19, 2020.
31. Daniel Giffney, "The Neurobiological Basis of Masochism: Why It Hurts So Good," *Trinity News*, April 24, 2019.
32. Orli Dahan, "Submission, Pain, and Pleasure: Considering an Evolutionary Hypothesis Concerning Sexual Masochism," *Psychology of Consciousness: Theory, Research, and Practice* 6, no. 4 (2019): 386–403.

Chapter 9: Keeping Pain in Mind

1. Donatella Marazziti, "Pain and Psychiatry: A Critical Analysis and Pharmacological Review," *Clinical Practice and Epistemology in Mental Health* (November 2006).
2. Catherine Q. Howe and Mark D. Sullivan, "The Missing 'P' in Pain Management: How the Current Opioid Epidemic Highlights the Need for Psychiatric Services in Chronic Pain Care," *General Hospital Psychiatry* (January 2014): 99–104.
3. Jenna Goesling, Lewei A. Lin, and Daniel J. Clauw, "Psychiatry and Pain Management: At the Intersection of Chronic Pain and Mental Health," *Current Psychology Reports* (March 2018).
4. Igor Elman, Jon-Kar Zubieta, and David Borsook, "Why It Is Important to Teach Pain to Psychiatrists," *JAMA Psychiatry Perspectives* (February 2021).
5. Elman, Zubieta, and Borsook, "Why It Is Important to Teach Pain to Psychiatrists."
6. The editors of Encyclopedia Britannica, "Lobotomy," Encyclopedia Britannica, last updated August 24, 2022, https://www.britannica.com/science/lobotomy.

7. Tanya Lewis, "Lobotomy: Definition, Procedure and History," LiveScience, August 28, 2018.
8. Lewis, "Lobotomy."
9. Lewis, "Lobotomy."
10. "Electroconvulsive Therapy," The American Psychiatric Association, July 2019, www.psychiatry.org/patients-families/ect.
11. John D. Otis and Douglas H. Hughes, "Psychiatry and Chronic Pain," *Psychiatric Times*, December 16, 2010.
12. Bernard Carey, "The Psychiatrist Will See You Online Now," *New York Times*, August 28, 2020, https://covidresponse.bidmcgiving.org/news/the-psychiatrist-will-see-you-online-now/.
13. Carey, "The Psychiatrist Will See You Online Now."
14. Carey, "The Psychiatrist Will See You Online Now."
15. "Minimum Insurance Benefits for Patients with Chronic Pain," American Academy of Pain Medicine, April 4, 2019.
16. Dennis Thompson Jr., "Will Insurance Cover Your Pain Treatment?" Everyday Health, March 9, 2020.
17. Thompson, "Will Insurance Cover Your Pain Treatment?"
18. Thompson, "Will Insurance Cover Your Pain Treatment?"
19. Paul Demko, "Health Plans Don't Want Patients on Opioids. So What Are They Doing for Pain?" *Politico*, February 12, 2019.
20. Demko, "Health Plans."
21. Demko, "Health Plans."
22. Joint Report of the Council on Medical Service and Council on Science and Public Health, Coverage for Chronic Pain Management, American Medical Association, www.ama-assn.org/sites/ama-assn.org/files/corp/media-browser/public/hod/a15-hod-joint-reports.pdf.
23. Peter Roy-Byrne, "Chronic Pain a Risk Factor for Suicide," *NEJM Journal Watch*, September 17, 2018.
24. Martin Cheatle, "Depression, Chronic Pain and Suicide by Overdose," *Pain Medicine* (June 2011): 43–48.
25. Ann-Marie D'arcy Sharpe, "Chronic Pain and Suicide," Pathways, February 5, 2020.
26. Cheatle, "Depression, Chronic Pain and Suicide by Overdose."
27. Melissa Wardlaw, "A Letter to Those Who Are Suicidal Because of Chronic Pain," *The Mighty*, January 10, 2020.
28. Kathleen Crowley, "Managing Depression: While Chronic Pain and Depression Can Go Hand in Hand, They Don't Have To," National Chronic Pain Outreach Association *Lifeline*, Winter 1996, https://walkforhealing.weebly.com/managing-chronic-pain-and-depression.html.
29. Robert Gatchel, Yuan B. Peng, Madelon L. Peters, Perry N. Fuchs, "The Biopsychosocial Approach to Chronic Pain: Scientific Advances and

Future Directions," *Psychological Bulletin* 133, no. 4 (August 2007): 581–624.
30. Linda Geddes,"Sufferers of Chronic Pain Have Long Been Told It's All in Their Head. We Now Know That's Wrong," *Guardian*, June 28, 2021, www.theguardian.com/australia-news/2021/jun/28/sufferers-of-chronic-pain-have-long-been-told-its-all-in-their-head-we-now-know-thats-wrong.
31. "Antidepressants: Another Weapon against Chronic Pain," Mayo Clinic, September 7, 2019, www.mayoclinic.org/pain-medications/art-20045647.
32. "Antidepressants: Another Weapon against Chronic Pain," Mayo Clinic.
33. Fava, Maurizio. "Somatic Symptoms, Depression, and Antidepressant Treatment," *Journal of Clinical Psychiatry* 63, no. 4 (2002): 305–7.
34. "Hurting Bodies and Suffering Minds Often Require the Same Treatment," Harvard Medical School, March 21, 2017, www.health.harvard.edu/mind-and-mood/depression-and-pain.
35. Aisha Morris Moultry and Ivy G. Poon, "The Use of Antidepressants for Chronic Pain," *U.S. Pharmacist* 34, no. 5 (May 2009).
36. "Anxiety," The Institute for Chronic Pain, last modified March 14, 2021, www.instituteforchronicpain.org/understanding-chronic-pain/complications/anxiety.
37. Yvette Terrie, "Medical Cannabis for Chronic Pain," *U.S. Pharmacist* 44, no. 3 (November 2020): 24–28.
38. Terrie, "Medical Cannabis."
39. Robert Dworkin, "Pain Insensitivity in Schizophrenia: A Neglected Phenomenon and Some Implications," *Schizophrenia Bulletin* 20, no. 2 (1994).
40. Douglas H. Hughes and John D. Otis, "Psychiatry and Chronic Pain," *Psychiatric Times* 27, no. 12 (December 2010), www.psychiatrictimes.com/view/psychiatry-and-chronic-pain.
41. Otis and Hughes, "Psychiatry and Chronic Pain."

Chapter 10: The Road Ahead

1. Saba Javed, Joey Hung, and Billy K. Huh, "Impact of COVID-19 on Chronic Pain Patients: A Pain Physician's Perspective," *Pain Management* (August 2020).
2. Daniel J. Clauw, "Considering the Potential for an Increase in Chronic Pain after the COVID-19 Pandemic," *Pain* 261, no. 8 (August 2020): 1694–97.
3. Brandon May, "Neuropathic Pain and COVID-19: Current Understanding and Future Directions," *Clinical Pain Advisor*, March 11, 2021.
4. "Coronavirus and the Nervous System," National Institute of Neurological Disorders and Stroke, www.ninds.nih.gov/current-research/coronavirus-and-ninds/coronavirus-and-nervous-system.

5. Linda Geddes, "Sufferers of Chronic Pain Have Long Been Told It's in Their Head. We Now Know That's Wrong," *Guardian*, June 28, 2021.
6. Clauw, "Considering the Potential for an Increase in Chronic Pain."
7. Geddes, "Sufferers of Chronic Pain."
8. Kai Karas, "The Social Threats of COVID-19 for People with Chronic Pain," *Pain* 161, no. 10 (October 2020): 2229–35.
9. Barbara Greenwood Dufour, "New Drugs for Chronic Pain: The Search Continues for Something Better Than Opioids," *Hospital News*, May 3, 2018, https://hospitalnews.com/new-drugs-chronic-pain-search-continues-something-better-opioids/.
10. Geddes, "Sufferers of Chronic Pain."
11. Akhtar Purvez, *Managing Chronic Pain in an Age of Addiction* (Lanham, MD: Rowman & Littlefield, 2018).
12. Purvez, *Managing Chronic Pain.*
13. Purvez, *Managing Chronic Pain.*
14. Moheb Costandi, "The Future of Chronic Pain," NEO-LIFE, May 6, 2021.
15. "Transcranial Focused Ultrasound Shows Promise to Treat Forms of Chronic Pain," *Medical Life Sciences*, August 16, 2021, www.news-medical.net/news/20210816/Transcranial-focused-ultrasound-shows-promise-to-treat-forms-of-chronic-pain.aspx.
16. Steven Richeimer and John Lee, "Genetic Testing in Pain Medicine: The Future Is Coming," *Practical Pain Management* 6, no. 8 (2016).

Index

About the Author

Akhtar Purvez, MD, author of *Managing Chronic Pain in an Age of Addiction*, is a researcher, interventional pain physician, and pain advocate. He is certified by the American Board of Pain Medicine, the American Board of Anesthesiology in Pain Medicine, and the American Board of Disability Analysts, and he has been a member of the Spine Intervention Society (SIS), the American Society of Regional Anesthesia and Pain Medicine (ASRA), and the American Medical Association (AMA). He was adjunct clinical professor at Lincoln Memorial University and has been involved in training physician assistants, medical students, residents, fellows, and other physicians.

Currently, he sits on the advisory board of the *Journal of Integrative Medicine and Public Health* and writes, lectures, and speaks on radio and TV about pain-related issues, supporting pain advocacy, pain policy, training, and research. He lives and practices in beautiful central Virginia.